ISBN 978-1-330-17652-8
PIBN 10045077


FB &c Ltd, Dalton House, 60 Windsor Avenue, London, SW19 2RR.
Company number 08720141. Registered in England and Wales.

# DEUCALION.

COLLECTED STUDIES

OF THE

*LAPSE OF WAVES, AND LIFE OF STONES.*

BY

JOHN RUSKIN, D.C.L., LL.D.,

HONORARY STUDENT OF CHRISTCHURCH, OXFORD; AND HONORARY FELLOW OF CORPUS CHRISTI COLLEGE, OXFORD.

ἐπειὴ μάλα πολλὰ μεταξὺ
οὔρεά τε σκιόεντα, θάλασσά τε ἠχήεσσα·

VOL. I.

NEW YORK:
JOHN WILEY & SONS,
15 ASTOR PLACE.
1886.

# CONTENTS OF VOL. I.

## CHAPTER IX.



## CHAPTER X.



## CHAPTER XI.



## CHAPTER XII.



## CHAPTER XIII.



## CHAPTER XIV.



---

# DEUCALION.

---

## INTRODUCTION.

BRANTWOOD, 13*th July*, 1875.

I HAVE been glancing lately at many biographies, and have been much struck by the number of deaths which occur between the ages of fifty and sixty, (and, for the most part, in the earlier half of the decade,) in cases where the brain has been much used emotionally: or perhaps it would be more accurate to say, where the heart, and the faculties of perception connected with it, have stimulated the brain-action. Supposing such excitement to be temperate, equable, and joyful, I have no doubt the tendency of it would be to prolong, rather than depress, the vital energies. But the emotions of indignation, grief, controversial anxiety and vanity, or hopeless, and therefore uncontending, scorn, are all of them as deadly to the body as poisonous air or polluted water; and when I reflect how much of the active part of my past life has been spent in these states,—and that what may remain to me of life can never more be in any other,—I begin to ask myself, with somewhat pressing arithmetic, how much time is likely to be left me, at the age of fifty-six, to complete the various designs for which, until past fiftv, I was merely collecting materials.

Of these materials, I have now enough by me for a most interesting (in my own opinion) history of fifteenth-century Florentine art, in six octavo volumes; an analysis of the Attic art of the fifth century B.C., in three volumes; an exhaustive history of northern thirteenth-century art, in ten volumes; a life of Turner, with analysis of modern landscape art, in four volumes; a life of Walter Scott, with analysis of modern epic art, in seven volumes; a life of Xenophon, with analysis of the general principles of Education, in ten volumes; a commentary on Hesiod, with final analysis of the principles of Political Economy, in nine volumes; and a general description of the geology and botany of the Alps, in twenty-four volumes.

Of these works, though all carefully projected, and some already in progress,—yet, allowing for the duties of my Professorship, possibly continuing at Oxford, and for the increasing correspondence relating to Fors Clavigera,—it does not seem to me, even in my most sanguine moments, now probable that I shall live to effect such conclusion as would be satisfactory to me; and I think it will therefore be only prudent, however humiliating, to throw together at once, out of the heap of loose stones collected for this many-towered city which I am not able to finish, such fragments of good marble as may perchance be useful to future builders; and to clear away, out of sight, the lime and other rubbish which I meant for mortar.

And because it is needful, for my health's sake, henceforward to do as far as possible what I find pleasure, or at least tranquillity, in doing, I am minded to collect first what I have done in geology and botany; for indeed, had it not been for grave mischance in earlier life, (partly consisting in the unlucky gift, from an affectionate friend, of Rogers' poems, as related in Fors Clavigera for August of this year,) my natural disposition for these sciences would certainly long ago have made me a leading member of the British Association for the Advancement of Science; or —who knows?—even raised me to the position which it was always the summit of my earthly ambition to attain, that of President of the Geological Society. For, indeed, I began when I was only twelve years old, a 'Mineralogical Dictionary,' intended to supersede everything done by Werner and Mohs, (and written in a shorthand composed of crystallographic signs now entirely unintelligible to me,)—and year by year have endeavoured, until very lately, to keep abreast with the rising tide of geological knowledge; sometimes even, I believe, pushing my way into little creeks in advance of the general wave. I am not careful to assert for myself the petty advantage of priority in discovering what, some day or other, somebody must certainly have discovered. But I think it due to my readers, that they may receive what real good there may be in these studies with franker confidence, to tell them that the first sun-portrait ever taken of the Matterhorn, (and as far as I know of any Swiss mountain whatever,) was taken by me in the year 1849; that the outlines, (drawn by measurement of angle,) given in 'Modern Painters' of the Cervin, and aiguilles of Chamouni, are at this day demonstrable by photography as the trustworthiest then in existence; that I was the first to point out, in my lecture given in the Royal Institution,* the real relation of the vertical cleavages to the stratification, in the limestone ranges belonging to the chalk formation in Savoy; and that my analysis of the structure of agates, ('Geological Magazine,') remains, even to the present day, the only one which has the slightest claim to accuracy of distinction, or completeness of arrangement. I propose therefore, if time be spared me, to collect, of these detached studies, or lectures, what seem to me deserving of preservation; together with the more carefully written chapters on geology and botany in the latter volumes of 'Modern Painters;' adding the memoranda I have still by me in manuscript, and such further illustrations as may occur to me on revision. Which fragmentary work,—trusting that among the flowers or stones let fall by other hands it may yet find service and life,—I have ventured to dedicate to Proserpina and Deucalion.

Why not rather to Eve, or at least to one of the wives of Lamech, and to Noah? asks, perhaps, the pious modern reader.

* Reported in the 'Journal de Genève,' date ascertainable, but of no consequence.

Because I think it well that the young student should first learn the myths of the betrayal and redemption, as the Spirit which moved on the face of the wide first waters, taught them to the heathen world. And because, in this power, Proserpine and Deucalion are at least as true as Eve or Noah; and all four together incomparably truer than the Darwinian Theory. And, in general, the reader may take it for a first principle, both in science and literature, that the feeblest myth is better than the strongest theory: the one recording a natural impression on the imaginations of great men, and of unpretending multitudes; the other, an unnatural exertion of the wits of little men, and half-wits of impertinent multitudes.

It chanced, this morning, as I sat down to finish my preface, that I had, for my introductory reading the fifth chapter of the second book of Esdras; in which, though often read carefully before, I had never enough noticed the curious verse, "Blood shall drop out of wood, and the stone shall give his voice, and the people shall be troubled." Of which verse, so far as I can gather the meaning from the context, and from the rest of the chapter, the intent is, that in the time spoken of by the prophet, which, if not our own, is one exactly corresponding to it, the deadness of men to all noble things shall be so great, that the sap of trees shall be more truly blood, in God's sight, than their hearts' blood; and the silence of men, in praise of all noble things, so great, that the stones shall cry out, in God's hearing, instead of their tongues; and the rattling of the shingle on the beach, and the roar of the rocks driven by the torrent, be truer Te Deum than the thunder of all their choirs. The writings of modern scientific prophets teach us to anticipate a day when even these lower voices shall be also silent; and leaf cease to wave, and stream to murmur, in the grasp of an eternal cold. But it may be, that rather out of the mouths of babes and sucklings a better peace may be promised to the redeemed Jerusalem; and the strewn branches, and low-laid stones, remain at rest at the gates of the city, built in unity with herself, and saying with her human voice, "My King cometh."

# CHAPTER I.

## THE ALPS AND JURA.

(*Part of a Lecture given in the Museum of Oxford, in October*, 1874.)

1. It is often now a question with me whether the persons who appointed me to this Professorship have been disappointed, or pleased, by the little pains I have hitherto taken to advance the study of landscape. That it is my own favourite branch of painting seemed to me a reason for caution in pressing it on your attention; and the range of art-practice which I have hitherto indicated for you, seems to me more properly connected with the higher branches of philosophical inquiry native to the University. But, as the second term of my Professorship will expire next year, and as I intend what remains of it to be chiefly employed in giving some account of the art of Florence and Umbria, it seemed to me proper, before entering on that higher subject, to set before you some of the facts respecting the great elements of landscape, which I first stated thirty years ago; arranging them now in such form as my farther study enables me to give them. I shall not, indeed, be able to do this in a course of spoken lec-

tures; nor do I wish to do so. Much of what I desire that you should notice is already stated, as well as I can do it, in 'Modern Painters;' and it would be waste of time to recast it in the form of address. But I should not feel justified in merely reading passages of my former writings to you from this chair; and will only ask your audience, here, of some additional matters, as, for instance, to-day, of some observations I have been making recently, in order to complete the account given in 'Modern Painters,' of the structure and aspect of the higher Alps.

2. Not that their structure—(let me repeat, once more, what I am well assured you will, in spite of my frequent assertion, find difficult to believe,)—not that their structure is any business of yours or mine, as students of practical art. All investigations of internal anatomy, whether in plants, rocks, or animals, are hurtful to the finest sensibilities and instincts of form. But very few of us have any such sensibilities to be injured; and that we may distinguish the excellent art which they have produced, we must, by duller processes, become cognizant of the facts. The Torso of the Vatican was not wrought by help from dissection; yet all its supreme qualities could only be explained by an anatomical master. And these drawings of the Alps by Turner are in landscape, what the Elgin marbles or the Torso are in sculpture. There is nothing else approaching them, or of their order. Turner made them before geology existed; but it is only by help of geology that I can prove their power.

3. I chanced, the other day, to take up a number of the 'Alpine Journal' (May, 1871,) in which there was a review by Mr. Leslie Stephen, of Mr. Whymper's 'Scrambles among the Alps,' in which it is said that "if the Alpine Club has done nothing else, it has taught us for the first time really to see the mountains." I have not the least idea whom Mr. Stephen means by 'us;' but I can assure him that mountains had been seen by several people before the nineteenth century; that both Hesiod and Pindar occasionally had eyes for Parnassus, Virgil for the Apennines, and Scott for the Grampians; and without speaking of Turner, or of any other accomplished artist, here is a little bit of old-fashioned Swiss drawing of the two Mythens, above the central town of Switzerland,* showing a degree of affection, intelligence, and tender observation, compared to which our modern enthusiasm is, at best, childish; and commonly also as shallow as it is vulgar.

4. Believe me, gentlemen, your power of seeing mountains cannot be developed either by your vanity, your curiosity, or your love of muscular exercise. It depends on the cultivation of the instrument of sight itself, and of the soul that uses it. As soon as you can see mountains rightly, you will see hills also, and valleys, with considerable interest; and a great many other things in Switzerland with which you are at present but poorly acquainted. The bluntness of your present capacity of ocular sensa

* In the Educational Series of my Oxford Schools.

tion is too surely proved by your being unable to enjoy any of the sweet lowland country, which is incomparably more beautiful than the summits of the central range, and which is meant to detain you, also, by displaying—if you have patience to observe them—the loveliest aspects of that central range itself, in its real majesty of proportion, and mystery of power.

5. For, gentlemen, little as you may think it, you can no more see the Alps from the Col du Géant, or the top of the Matterhorn, than the pastoral scenery of Switzerland from the railroad carriage. If you want to see the skeletons of the Alps, you may go to Zermatt or Chamouni; but if you want to see the body and soul of the Alps, you must stay awhile among the Jura, and in the Bernese plain. And, in general, the way to see mountains, is to take a knapsack and a walking-stick; leave alpenstocks to be flourished in each other's faces, and between one another's legs, by Cook's tourists; and try to find some companionship in yourself with yourself; and not to be dependent for your good cheer either on the gossip of the table-d'hôte, or the hail-fellow and well met, hearty though it be, of even the pleasantest of celebrated guides.

6. Whether, however, you think it necessary or not, for true sight of the Alps, to stay awhile among the Jura or in the Bernese fields, very certainly, for understanding, or questioning, of the Alps, it is wholly necessary to do so. If you look back to the lecture, which I gave as the fourth of my inaugural series, on the Relation of Art to Use, you

will see it stated, as a grave matter of reproach to the modern traveller, that, crossing the great plain of Switzerland nearly every summer, he never thinks of inquiring why it is a plain, and why the mountains to the south of it are mountains.

7. For solution of which, as it appears to me, not unnatural inquiry, all of you, who have taken any interest in geology whatever, must recognize the importance of studying the calcareous ranges which form the outlying steps of the Alps on the north; and which, in the lecture just referred to, I requested you to examine for their crag scenery, markedly developed in the Stockhorn, Pilate, and Sentis of Appenzell. The arrangements of strata in that great calcareous belt give the main clue to the mode of elevation of the central chain, the relations of the rocks over the entire breadth of North Switzerland being, roughly, as in this first section:

FIG. 1.

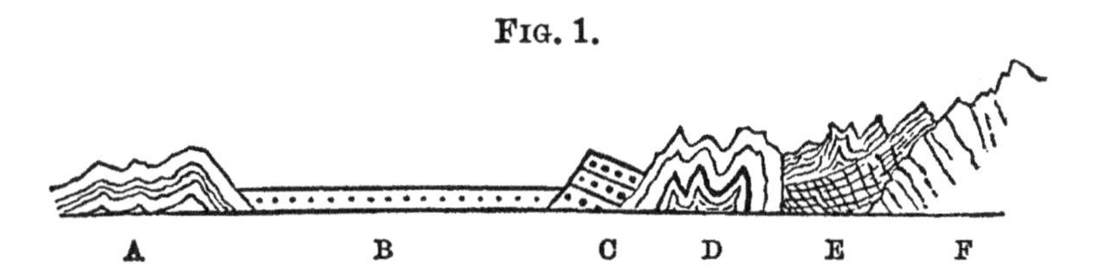


A. Jura limestones, moderately undulating in the successive chains of Jura.

B. Sandstones of the great Swiss plain.

C. Pebble breccias of the first ranges of Alpine hills.

D. Chalk formations violently contorted, forming the rock scenery of which I have just spoken.

E. Metamorphic rocks lifted by the central Alps.

F. Central gneissic or granitic mass, narrow in Mont Blanc, but of enormous extent southward from St. Gothard.

8. Now you may, for first grasp of our subject, imagine these several formations all fluted longitudinally, like a Gothic moulding, thus forming a series of ridges and valleys parallel to the Alps;—such as the valley of Chamouni, the Simmenthal, and the great vale containing the lakes of Thun and Brienz; to which longitudinal valleys we now obtain access through gorges or defiles, for the most part cut across the formations, and giving geological sections all the way from the centres of the Alps to the plain.

9. Get this first notion very simply and massively set in your thoughts. Longitudinal valleys, parallel with the beds; more or less extended and soft in contour, and often occupied by lakes. Cross defiles like that of Lanterbrunnen, the Via Mala, and the defile of Gondo; cut down across the beds, and traversed by torrents, but rarely occupied by lakes. The bay of Uri is the only perfect instance in Switzerland of a portion of lake in a diametrically cross valley; the crossing arms of the lake Lucerne mark the exactly rectangular schism of the forces; the main direction being that of the lakes of Kussnacht and Alpnacht, carried on through those of Sarnen and Lungern, and across the low intervening ridge of the Brunig, joining the depressions of Brienz and Thun; of which last lake the

lower reach, however, is obliquely transverse. Forty miles of the Lago Maggiore, or, including the portion of lake now filled by delta, fifty, from Baveno to Bellinzona, are in the longitudinal valley which continues to the St. Bernardino: and the entire length of the lake of Como is the continuation of the great lateral Valtelline.

10. Now such structure of parallel valley and cross defile would be intelligible enough, if it were confined to the lateral stratified ranges. But, as you are well aware, the two most notable longitudinal valleys in the Alps are cut right along the heart of their central gneissic chain; how much by dividing forces in the rocks themselves, and how much by the sources of the two great rivers of France and Germany, there will yet be debate among geologists for many a day to come. For us, let the facts at least be clear; the questions definite; but all debate declined.

11. All lakes among the Alps, except the little green pool of Lungern, and a few small tarns on the cols, are quite at the bottom of the hills. We are so accustomed to this condition, that we never think of it as singular. But in its unexceptional character, it is extremely singular. How comes it to pass, think you, that through all that wilderness of mountain—raised, in the main mass of it, some six thousand feet above the sea, so that there is no col lower,—there is not a single hollow shut in so as to stay the streams of it;—that no valley is ever barred across by a ridge which can keep so much as ten feet of water

calm above it,—that every such ridge that once existed has been cut through, so as to let the stream escape?

I put this question in passing; we will return to it: let me first ask you to examine the broad relations of the beds that are cut through. My typical section, Fig. 1, is stringently simple; it must be much enriched and modified to fit any locality; but in the main conditions it is applicable to the entire north side of the Alps, from Annecy to St. Gall.

12. You have first—(I read from left to right, or north to south, being obliged to do so because all Studer's sections are thus taken)—this mass of yellow limestone, called of the Jura, from its development in that chain; but forming an immense tract of the surface of France also; and, as you well know, this our city of Oxford stands on one of its softer beds, and is chiefly built of it. We may, I think, without entering any forbidden region of theory, assume that this Jura limestone extends under the plain of Switzerland, to reappear where we again find it on the flanks of the great range; where on the top of it the beds drawn with fine lines in my section correspond generally to the date of our English chalk, though they are far from white in the Alps. Curiously adjusted to the chalk beds, rather than superimposed, we have these notable masses of pebble breccia, which bound the sandstones of the great Swiss plain.

13. I have drawn that portion of the section a little more boldly in projection, to remind you of the great

Rigi promontory; and of the main direction of the slope of these beds, with their backs to the Alps, and their escarpments to the plain. Both these points are of curious importance. Have you ever considered the reason of the fall of the Rossberg, the most impressive physical catastrophe that has chanced in Europe in modern times? Few mountains in Switzerland looked safer. It was of inconsiderable height, of very moderate steepness; but its beds lay perfectly straight, and that over so large a space, that when the clay between two of them got softened by rain, one slipped off the other. Now this mathematical straightness is characteristic of these pebble beds,—not universal in them, but characteristic of them, and of them only. The limestones underneath are usually, as you see in this section, violently contorted; if not contorted, they are at least so irregular in the bedding that you can't in general find a surface of a furlong square which will not either by its depression, or projection, catch and notch into the one above it, so as to prevent its sliding. Also the limestones are continually torn, or split, across the beds. But the breccias, though in many places they suffer decomposition, are curiously free from fissures and rents. The hillside remains unshattered unless it comes down in a mass. But their straight bedding, as compared with the twisted limestone, is the notablest point in them; and see how very many difficulties are gathered in the difference. The crushed masses of limestone are supposed to have been wrinkled together by the lateral thrust of the emerging protogines; and these pebble beds to have been raised into a gable, or broken into a series of colossal fragments set over each other like tiles, all along the south shore of the Swiss plain, by the same lateral thrust; nay, "though we may leave in doubt," says Studer, "by what cause the folded forms of the Jura may have been pushed back, there yet remains to us, for the explanation of this gabled form of the Nagelfluh, hardly any other choice than to adopt the opinion of a lateral pressure communicated by the Alps to the tertiary bottom. We have often found in the outer limestone chains themselves clear evidence of a pressure going out from the inner Alps; and the pushing of the older over the younger formations along the flank of the limestone hills, leaves hardly any other opinion possible."

14. But if these pebble beds have been heaved up by the same lateral thrust, how is it that a force which can

FIG. 2.

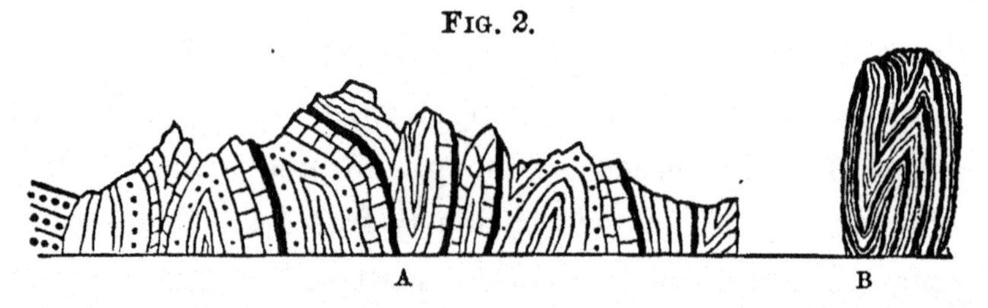


bend limestone like leather, cannot crush anywhere, these pebble beds into the least confusion? Consider the scale on which operations are carried on, and the forces of which this sentence of Studer's so serenely assumes the action. Here, A, Fig. 2, is his section of the High Sentis of Appenzell, of which the height is at least, in the parts thus bent, 6,000 feet. And here, B, Fig. 2, are some sheets of paper crushed together by my friend Mr. Henry Woodward, from a length of four inches, into what you see; the High Sentis, exactly resembles these, and seems to consist of four miles of limestone similarly crushed into one. Seems, I say, remember: I never theorize, I give you the facts only. The beds *do* go up and down like this: that they have been crushed together, it is Mr. Studer who says or supposes; I can't go so far; nevertheless, I admit that he appears to be right, and I believe he is right; only don't be positive about it, and don't debate; but think of it, and examine.

15. Suppose, then, you have a bed of rocks, four miles long by a mile thick, to be crushed laterally into the space of a mile. It may be done, supposing the mass not to be reducible in bulk, in two ways: you may either crush it up into folds, as I crush these pieces of cloth; or you may break it into bits, and shuffle them over one another like cards. Now, Mr. Studer, and our geologists in general, believe the first of these operations to have taken place with the limestones, and the second with the breccias. They are, as I say, very probably right: only just consider what is involved in the notion of shuffling up your breccias like a pack of cards, and folding up your limestones like a length of silk which a dexterous draper's shopman is persuading a young lady to put ten times as much of into her gown as is wanted for it! Think, I say, what is

involved in the notion. That you may shuffle your pebble beds, you must have them strong and well knit. Then what sort of force must you have to break and to heave them? Do but try the force required to break so much as a captain's biscuit by a slow push,—it is the illustration I gave long ago in 'Modern Painters,'—and then fancy the results of such fracturing power on a bed of conglomerate two thousand feet thick! And here is indeed a very charming bookbinder's pattern, produced by my friend in crushed paper, and the length of silk produces lovely results in these arrangements à la Paul Veronese. But when you have the cliffs of the Diablerets, or the Dent du Midi of Bex, to deal with; and have to fold *them* up similarly, do you mean to fold your two-thousand-feet-thick Jura limestone in a brittle state, or a ductile one? If brittle, won't it smash? If ductile, won't it squeeze? Yet your whole mountain theory proceeds on the assumption that it has neither broken nor been compressed,—more than the folds of silk or coils of paper.

16. You most of you have been upon the lake of Thun. You have been at least carried up and down it in a steamer; you smoked over it meanwhile, and countenanced the Frenchmen and Germans who were spitting into it. The steamer carried you all the length of it in half an hour; you looked at the Jungfrau and Blumlis Alp, probably, for five minutes, if it was a fine day; then took to your papers, and read the last news of the Tichborne case; then you lounged about,—thought it a nuisance that the

steamer couldn't take you up in twenty minutes, instead of half an hour; then you got into a row about your luggage at Neuhaus; and all that you recollect afterwards is that lunch where you met the so-and-sos at Interlaken.

17. Well, we used to do it differently in old times. Look here;—this * is the quay at Neuhaus, with its then travelling arrangements. A flat-bottomed boat, little better than a punt;—a fat Swiss girl with her schatz, or her father, to row it; oars made of a board tied to a pole: and so one paddled along over the clear water, in and out among the bays and villages, for half a day of pleasant life. And one knew something about the lake, ever after, if one had a head with eyes in it.

It is just possible, however, that some of you also who have been learning to see the Alps in your new fashion, may remember that the north side of the lake of Thun consists, first, next Thun, of a series of low green hills, with brown cliffs here and there among the pines; and that above them, just after passing Oberhofen, rears up suddenly a great precipice, with its flank to the lake, and the winding wall of it prolonged upwards, far to the north, losing itself, if the day is fine, in faint tawny crests of rock among the distant blue; and if stormy, in wreaths of more than commonly torn and fantastic cloud.

18. To form the top of that peak on the north side of

* Turner's first study of the Lake of Thun, 1803.

the lake of Thun, you have to imagine forces which have taken—say, the whole of the North Foreland, with Dover castle on it, and have folded it upside-down on the top of the parade at Margate,—then swept up Whitstable oyster-beds, and put them on the bottom of Dover cliffs turned topsy-turvy,—and then wrung the whole round like a wet towel, till it is as close and hard as it will knit;—such is the beginning of the operations which have produced the lateral masses of the higher Alps.

19. Next to these, you have the great sculptural force, which gave them, approximately, their present forms,—which let out all the lake waters above a certain level,—which cut the gorge of the Devil's Bridge—of the Via Mala—of Gondo—of the valley of Cluse;—which let out the Rhone at St. Maurice, the Ticino at Faido, and shaped all the vast ravines which make the flanks of the great mountains awful.

20. Then, finally, you have the ram, torrent, and glacier of human days.

Of whose action, briefly, this is the sum.

Over all the high surfaces, disintegration—melting away—diffusion—loss of height and terror.

In the ravines,—whether occupied by torrent or glacier,—gradual incumbrance by materials falling from above; choking up of their beds by silt—by moraine—by continual advances of washed slopes on their flanks: here and there, only, exceptional conditions occur in which a river is still continuing feebly the ancient cleaving action, and cutting its ravine deeper, or cutting it back.

Fix this idea thoroughly in your minds. Since the valley of Lauterbrunnen existed for human eyes,—or its pastures for the food of flocks,—it has not been cut deeper, but partially filled up by its torrents. The town of Interlachen stands where there was once lake,—and the long slopes of grassy sward on the north of it, stand where once was precipice. Slowly,—almost with infinite slowness,—the declining and encumbering action takes place; but incessantly, and,—as far as our experience reaches,–irredeemably.

21. Now I have touched in this lecture briefly on the theories respecting the elevation of the Alps, because I want to show you how uncertain and unsatisfactory they still remain. For our own work, we must waste no time on them; we must begin where all theory ceases; and where observation becomes possible,—that is to say, with the forms which the Alps have actually retained while men have dwelt among them, and on which we can trace the progress, or the power, of existing conditions of minor change. Such change has lately affected, and with grievous deterioration, the outline of the highest mountain of Europe, with that of its beautiful supporting buttresses,—the aiguille de Bionassay. I do not care, and I want you not to care,—how crest or aiguille was lifted, or where its materials came from, or how much bigger it was once. I do care that you should know, and I will endeavour in

these following pages securely to show you, in strength and beauty of form it has actually stood man was man, and what subtle modifications of asp majesties of contour, it still suffers from the rain beat upon it, or owes to the snows that rest.

# CHAPTER II.

## THE THREE ÆRAS.

(*Part of a Lecture given at the London Institution in March*, 1875, *with added pieces from Lectures in Oxford.*)

1. We are now, so many of us, some restlessly and some wisely, in the habit of spending our evenings abroad, that I do not know if any book exists to occupy the place of one classical in my early days, called 'Evenings at Home.' It contained, among many well-written lessons, one, under the title of 'Eyes and No Eyes,' which some of my older hearers may remember, and which I should myself be sorry to forget. For if such a book were to be written in these days, I suppose the title and the moral of the story would both be changed; and, instead of 'Eyes and No Eyes,' the tale would be called 'Microscopes and No Microscopes.' For I observe that the prevailing habit of learned men is now to take interest only in objects which cannot be seen without the aid of instruments; and I believe many of my learned friends, if they were permitted to make themselves, to their own liking, instead of suffering the slow process of selective development, would give

themselves heads like wasps', with three microscopic eyes in the middle of their foreheads, and two ears at the ends of their antennae.

2. It is the fashion, in modern days, to say that Pope was no poet. Probably our schoolboys also, think Horace none. They have each, nevertheless, built for themselves a monument of enduring wisdom; and all the temptations and errors of our own day, in the narrow sphere of lenticular curiosity, were anticipated by Pope, and rebuked, in one couplet:

> "Why has not man a microscopic eye?
> For this plain reason,—Man is not a fly."

While the nobler following lines,

> "Say, what avail, were finer optics given
> To inspect a mite, not comprehend the heaven?"

only fall short of the truth of our present dulness, in that we inspect heaven itself, without understanding it.

3. In old times, then, it was not thought necessary for human creatures to know either the infinitely little, or the infinitely distant; nor either to see, or feel, by artificial help. Old English people used to say they perceived things with their five—or it may be, in a hurry, they would say, their seven, *senses*; and that word 'sense' became, and for ever must remain, classical English, derived from classical Latin, in both languages signifying, not only the bodily sense, but the moral one. If a man heard,

saw, and tasted rightly, we used to say he had his bodily senses perfect. If he judged, wished, and felt rightly, we used to say he had his moral senses perfect, or was a man 'in his senses.' And we were then able to speak precise truth respecting both matter and morality; and if we heard any one saying clearly absurd things,—as, for instance, that human creatures were automata,—we used to say they were out of their 'senses,' and were talking non-'sense.'

Whereas, in modern days, by substituting analysis for sense in morals, and chemistry for sense in matter, we have literally blinded ourselves to the essential qualities of both matter and morals; and are entirely incapable of understanding what is meant by the description given us, in a book we once honoured, of men who "by reason of use, have their *senses* exercised to discern both good and evil."

4. And still, with increasingly evil results to all of us, the separation is every day widening between the man of science and the artist—in that, whether painter, sculptor, or musician, the latter is pre-eminently a person who sees with his Eyes, hears with his Ears, and labours with his Body, as God constructed them; and who, in using instruments, limits himself to those which convey or communicate his human power, while he rejects all that increase it. Titian would refuse to quicken his touch by electricity; and Michael Angelo to substitute a steam hammer for his mallet. Such men not only do not desire, they impera

tively and scornfully refuse, either the force, or the information, which are beyond the scope of the flesh and the senses of humanity. And it is at once the wisdom, the honour, and the peace, of the Masters both of painting and literature, that they rejoice in the strength, and rest in the knowledge, which are granted to active and disciplined life; and are more and more sure, every day, of the wisdom of the Maker in setting such measure to their being; and more and more satisfied, in their sight and their audit of Nature, that "the hearing ear, and the seeing eye,—the Lord hath made even both of them."

5. This evening, therefore, I venture to address you speaking limitedly as an artist; but, therefore, I think, with a definite advantage in having been trained to the use of my eyes and senses, as my chief means of observation: and I shall try to show you things which with your own eyes you may any day see, and with your own common sense, if it please you to trust it, account for.

Things which you may see, I repeat; not which you might perhaps have seen, if you had been born when you were not born; nor which you might perhaps in future see, if you were alive when you will be dead. But what, in the span of earth, and space of time, allotted to you, may be seen with your human eyes, if you learn to use them.

And this limitation has, with respect to our present subject, a particular significance, which I must explain to you before entering on the main matter of it.

6. No one more honours the past labour—no one more regrets the present rest—of the late Sir Charles Lyell, than his scholar, who speaks to you. But his great theorem of the constancy and power of existing phenomena was only in measure proved,—in a larger measure disputable; and in the broadest bearings of it, entirely false. Pardon me if I spend no time in qualifications, references, or apologies, but state clearly to you what Sir Charles Lyell's work itself enables us now to perceive of the truth. There are, broadly, three great demonstrable periods of the Earth's history. That in which it was crystallized; that in which it was sculptured; and that in which it is now being unsculptured, or deformed. These three periods interlace with each other, and gradate into each other—as the periods of human life do. Something dies in the child on the day that it is born,—something is born in the man on the day that he dies: nevertheless, his life is broadly divided into youth, strength, and decrepitude. In such clear sense, the Earth has its three ages: of their length we know as yet nothing, except that it has been greater than any man had imagined.

7. (THE FIRST PERIOD.)—But there was a period, or a succession of periods, during which the rocks which are now hard were soft; and in which, out of entirely different positions, and under entirely different conditions from any now existing or describable, the masses, of which the mountains you now see are made, were lifted,

and hardened, in the positions they now occupy, though in what forms we can now no more guess than we can the original outline of the block from the existing statue.

8. (THE SECOND PERIOD.)—Then, out of those raised masses, more or less in lines compliant with their crystalline structure, the mountains we now see were hewn, or worn, during the second period, by forces for the most part differing both in mode and violence from any now in operation, but the result of which was to bring the surface of the earth into a form approximately that which it has possessed as far as the records of human history extend. The Ararat of Moses's time, the Olympus and Ida of Homer's, are practically the same mountains now, that they were then.

9. (THE THIRD PERIOD.)—Not, however, without some calculable, though superficial, change, and that change, one of steady degradation. For in the third, or historical period, the valleys excavated in the second period are being filled up, and the mountains, hewn in the second period, worn or ruined down. In the second æra the valley of the Rhone was being cut deeper every day; now it is every day being filled up with gravel. In the second æra, the scars of Derbyshire and Yorkshire were cut white and steep; now they are being darkened by vegetation, and crumbled by frost. You cannot, I repeat, separate the periods with precision; but, in their characters, they are as distinct as youth from age.

10. The features of mountain form, to which during my

own life I have exclusively directed my study, and which I endeavour to bring before the notice of my pupils in Oxford, are exclusively those produced by existing forces, on mountains whose form and substance have not been materially changed during the historical period.

For familiar example, take the rocks of Edinburgh Castle, and Salisbury Craig. Of course we know that they are both basaltic, and must once have been hot. But I do not myself care in the least what happened to them till they were cold.* They have both been cold at least long-

---

* More curious persons, who *are* interested in their earlier condition, will find a valuable paper by Mr. J. W. Judd, in the quarterly 'Journal of the Geological Society,' May, 1875; very successfully, it seems to me, demolishing all former theories on the subject, which the author thus sums, at p. 135.

"The series of events which we are thus required to believe took place in this district is therefore as follows:—

A. At the point where the Arthur's Seat group of hills now rises, a series of volcanic eruptions occurred during the Lower Calciferous Sandstone period, commencing with the emission of basaltic lavas, and ending with that of porphyrites.

B. An interval of such enormous duration supervened as to admit of—

*a.* The deposition of at least 3,000 feet of Carboniferous strata.

*b.* The bending of all the rocks of the district into a series of great anticlinal and synclinal folds.

*c.* The removal of every vestige of the 3,000 feet of strata by denudation.

C. The outburst, after this vast interval, of a second series of volcanic eruptions upon the *identical site* of the former ones, presenting in its

er than young Harry Percy's spur; and, since they were last brought out of the oven, in the shape which, approximately, they still retain, with a hollow beneath one of them, which, for aught I know, or care, may have been cut by a glacier out of white-hot lava, but assuredly at last got itself filled with pure, sweet, cold water, and called, in Lowland Scotch, the 'Nor' Loch;'—since the time, I say, when the basalt, above, became hard, and the lake beneath, drinkable, I am desirous to examine with you what effect the winter's frost and summer's rain have had on the crags and their hollows; how far the 'Kittle nine steps' under the castle-walls, or the firm slope and cresting precipice above the dark ghost of Holyrood, are enduring or departing forms; and how long, unless the young engineers of New Edinburgh blast the incumbrance away, the departing mists of dawn may each day reveal the form, unchanged, of the Rock which was the strength of their Fathers.

11. Unchanged, or so softly modified that eye can scarcely trace, or memory measure, the work of time. Have you ever practically endeavoured to estimate the alterations of form in any hard rocks known to you, during the

---

succession, of events *precisely the same sequence*, and resulting in the production of rocks of *totally undistinguishable character*.

Are we not entitled to regard the demand for the admission of such a series of extraordinary accidents as evidence of the *antecedent improbability* of the theory? And when we find that all attempts to suggest a period for the supposed second series of outbursts have successively failed, do not the difficulties of the hypothesis appear to be overwhelming?"

course of your own lives? You have all heard, a thousand times over, the common statements of the school of Sir Charles Lyell. You know all about alluviums and gravels; and what torrents do, and what rivers do, and what ocean currents do; and when you see a muddy stream coming down in a flood, or even the yellow gutter more than usually rampant by the roadside in a thunder shower, you think, of course, that all the forms of the Alps are to be accounted for by aqueous erosion, and that it's a wonder any Alps are still left. Well—any of you who have fished the pools of Scottish or a Welsh stream,—have you ever thought of asking an old keeper how much deeper they had got to be, while his hairs were silvering? Do you suppose he wouldn't laugh in your face?

There are some sitting here, I think, who must have themselves fished, for more than one summer, years ago, in Dove or Derwent,—in Tweed or Teviot. Can any of you tell me a single pool, even in the limestone or sandstone, where you could spear a salmon then, and can't reach one now—(providing always the wretches of manufacturers have left you one to be speared, or water that you can see through)? Do you know so much as a single rivulet of clear water which has cut away a visible half-inch of Highland rock, to your own knowledge, in your own day? You have seen whole banks, whole fields washed away; and the rocks exposed beneath? Yes, of course you have; and so have I. The rains wash the loose earth about everywhere, in any masses that they chance to catch

—loose earth, or loose rock. But yonder little rifted well in the native whinstone by the sheepfold,—did the gray shepherd not put his lips to the same ledge of it, to drink —when he and you were boys together?

12. 'But Niagara, and the Delta of the Ganges—and —all the rest of it?' Well, of course a monstrous mass of continental drainage, like Niagara, *will* wash down a piece of crag once in fifty years, (but only that, if it's rotten below;) and tropical rains will eat the end off a bank of slime and alligators,—and spread it out lower down. But does any Scotchman know a change in the Fall of Fyers? —any Yorkshireman in the Force of Tees?

Except of choking up, it may be—not of cutting down. It is true, at the side of every stream you see the places in the rocks hollowed by the eddies. I suppose the eddies go on at their own rate. But I simply ask, Has any human being ever known a stream, in hard rock, cut its bed an inch deeper down at a given spot?

13. I can look back, myself, now pretty nearly, I am sorry to say, half a century, and recognize no change whatever in any of my old dabbling-places; but that some stones are mossier, and the streams usually dirtier,—the Derwent above Keswick, for example.

'But denudation does go on, somehow: one sees the whole glen is shaped by it?' Yes, but not by the *stream*. The stream only sweeps down the loose stones; frost and chemical change are the powers that loosen them. I have indeed not known one of my dabbling-places changed in fifty years. But I have known the éboulement under the Rochers des Fyz, which filled the Lac de Chêde; I passed through the valley of Cluse a night after some two or three thousand tons of limestone came off the cliffs of Maglans—burying the road and field beside it. I have seen half a village buried by a landslip, and its people killed, under Monte St. Angelo, above Amalfi. I have seen the lower lake of Llanberis destroyed, merely by artificial slate quarries; and the Waterhead of Coniston seriously diminished in purity and healthy flow of current by the débris of its copper mines. These are all cases, you will observe, of degradation; diminishing majesty in the mountain, and diminishing depth in the valley, or pools of its waters. I cannot name a single spot in which, during my lifetime spent among the mountains, I have seen a peak made grander, a watercourse cut deeper, or a mountain pool made larger and purer.

14. I am almost surprised, myself, as I write these words, at the strength which, on reflection, I am able to give to my assertion. For, even till I began to write these very pages, and was forced to collect my thoughts, I remained under the easily adopted impression, that, at least among soft earthy eminences, the rivers were still cutting out their beds. And it is not so at all. There are indeed banks here and there which they visibly remove; but whatever they sweep down from one side, they sweep up on the other, and extend a promontory of land for every shelf they undermine: and as for those radiating fibrous

valleys in the Apennines, and such other hills, which look symmetrically shaped by streams,—they are not lines of trench from below, but lines of wash or slip from above: they are the natural wear and tear of the surface, directed indeed in easiest descent by the bias of the stream, but not dragged down by its grasp. In every one of those ravines the water is being choked up to a higher level; it is not gnawing down to a lower. So that, I repeat, earnestly, their chasms being choked below, and their precipices shattered above, all mountain forms are suffering a deliquescent and corroding change,—not a sculpturesque or anatomizing change. All character is being gradually effaced; all crooked places made straight,—all rough places, plain; and among these various agencies, not of *e*rosion, but *cor*rosion, none are so distinct as that of the glacier, in filling up, not cutting deeper, the channel it fills; and in rounding and smoothing, but never sculpturing, the rocks over which it passes.

In this fragmentary collection of former work, now patched and darned into serviceableness, I cannot finish my chapters with the ornamental fringes I used to twine for them; nor even say, by any means, all I have in my mind on the matters they treat of: in the present case, however, the reader will find an elucidatory postscript added at the close of the fourth chapter, which he had perhaps better glance over before beginning the third.

# CHAPTER III.

## OF ICE-CREAM.

*(Continuation of Lecture delivered at London Institution, with added Illustrations from Lectures at Oxford.)*

1. The statement at the close of the last chapter, doubtless surprising and incredible to many of my readers must, before I reinforce it, be explained as referring only to glaciers visible, at this day, in temperate regions. For of formerly deep and continuous tropical ice, or of existing Arctic ice, and their movements, or powers, I know, and therefore say, nothing.* But of the visible glaciers

* The following passage, quoted in the 'Geological Magazine' for June of this year, by Mr. Clifton Ward, of Keswick, from a letter of Professor Sedgwick's, dated May 24th, 1842, is of extreme value; and Mr. Ward's following comments are most reasonable and just:—

"No one will, I trust, be so bold as to affirm that an uninterrupted glacier could ever have extended from Shap Fells to the coast of Holderness, and borne along the blocks of granite through the whole distance, without any help from the floating power of water. The supposition involves difficulties tenfold greater than are implied in the phenomenon it pretends to account for. The glaciers descending through the valleys of the higher Alps have an enormous transporting power: but there is no such power in a great sheet of ice expanded over a country without mountains, and at a nearly dead level."

couched upon the visible Alps, two great facts are very clearly ascertainable, which, in my lecture at the London Institution, I asserted in their simplicity, as follows:—

2. The first great fact to be recognized concerning them is that they are *Fluid* bodies. Sluggishly fluid, indeed, but definitely and completely so; and therefore, they do not scramble down, nor tumble down, nor crawl down, nor slip down; but *flow* down. They do not move like leeches, nor like caterpillars, nor like stones, but like, what they are made of, water.

---

The difficulties involved in the theories of Messrs. Croll, Belt, Goodchild, and others of the same extreme school, certainly press upon me —and I think I may say also upon others of my colleagues—increasingly, as the country becomes more and more familiar in its features. It is indeed a most startling thought, as one stands upon the eastern borders of the Lake-mountains, to fancy the ice from the Scotch hills stalking boldly across the Solway, marching steadily up the Eden Valley, and persuading some of the ice from Shap to join it on an excursion over Stainmoor, and bring its boulders with it.

The outlying northern parts of the Lake-district, and the flat country beyond, have indeed been ravished in many a raid by our Scotch neighbours, but it is a question whether, in glacial times, the Cumbrian mountains and Pennine chain had not strength in their protruding icy arms to keep at a distance the ice proceeding from the district of the southern uplands, the mountains of which are not *superior* in elevation. Let us hope that the careful geological observations which will doubtless be made in the forthcoming *scientific* Arctic Expedition will throw much new light on our past glacial period.

J. CLIFTON WARD.

KESWICK, *April 26th*, 1875.

That is the main fact in their state, and progress, on which all their great phenomena depend.

Fact first discovered and proved by Professor James Forbes, of Edinburgh, in the year 1842, to the astonishment of all the glacier theorists of his time;—fact strenuously denied, disguised, or confusedly and partially apprehended, by all of the glacier theorists of subsequent times, down to our own day; else there had been no need for me to tell it you again to-night.

3. The second fact of which I have to assure you is partly, I believe, new to geologists, and therefore may be of some farther interest to you because of its novelty, though I do not myself care a grain of moraine-dust for the newness of things; but rather for their oldness; and wonder more willingly at what my father and grandfather thought wonderful, (as, for instance, that the sun should rise, or a seed grow,) than at any newly-discovered marvel. Nor do I know, any more than I care, whether this that I have to tell you be new or not; but I did not absolutely *know* it myself, until lately; for though I had ventured with some boldness to assert it as a consequence of other facts, I had never been under the bottom of a glacier to look. But, last summer, I was able to cross the dry bed of a glacier, which I had seen flowing, two hundred feet deep, over the same spot, forty years ago. And there I saw, what before I had suspected, that modern glaciers, like modern rivers, were not cutting their beds deeper, but filling them up. These, then, are the two facts I

wish to lay distinctly before you this evening,—first that glaciers are fluent; and, secondly, that they are filling up their beds, not cutting them deeper.

4. (I.) Glaciers are fluent; slowly, like lava, but distinctly.

And now I must ask you not to disturb yourselves, as I speak, with bye-thoughts about 'the theory of regelation.' It is very interesting to know that if you put two pieces of ice together, they will stick together; let good Professor Faraday have all the credit of showing us that; and the human race in general, the discredit of not having known so much as that, about the substance they have skated upon, dropped through, and eat any quantity of tons of—these two or three thousand years.

It was left, nevertheless, for Mr. Faraday to show them that two pieces of ice will stick together when they touch —as two pieces of hot glass will. But the capacity of ice for sticking together no more accounts for the making of a glacier, than the capacity of glass for sticking together accounts for the making of a bottle. The mysteries of crystalline vitrification, indeed, present endless entertainment to the scientific inquirer; but by no theory of vitrification can he explain to us how the bottle was made narrow at the neck, or dishonestly vacant at the bottom. Those conditions of it are to be explained only by the study of the centrifugal and moral powers to which it has been submitted.

5. In like manner, I do not doubt but that wonderful

phenomena of congelation, regelation, degelation, and gelation pure without preposition, take place whenever a schoolboy makes a snowball; and that miraculously rapid changes in the structure and temperature of the particles accompany the experiment of producing a star with it on an old gentleman's back. But the principal conditions of either operation are still entirely dynamic. To make your snowball hard, you must squeeze it hard; and its expansion on the recipient surface is owing to a lateral diversion of the impelling forces, and not to its regelatic properties.

6. Our first business, then, in studying a glacier, is to consider the mode of its original deposition, and the large forces of pressure and fusion brought to bear on it, with their necessary consequences on such a substance as we practically know snow to be,—a powder, ductile by wind, compressible by weight; diminishing by thaw, and hardening by time and frost; a thing which sticks to rough ground, and slips on smooth; which clings to the branch of a tree, and slides on a slated roof.

7. Let us suppose, then, to begin with, a volcanic cone in which the crater has been filled, and the temperature cooled, and which is now exposed to its first season of glacial agencies. Then let Plate 1, Fig. 1, represent this mountain, with part of the plans at its foot under an equally distributed depth of a first winter's snow, and place the level of perpetual snow at any point you like—for simplicity's sake, I put it halfway up the cone.

Below this snow-line, all snow disappears in summer; but above it, the higher we ascend, the more of course we find remaining. It is quite wonderful how few feet in elevation make observable difference in the quantity of snow that will lie. This last winter, in crossing the moors of the peak of Derbyshire, I found, on the higher masses of them, that ascents certainly not greater than that at Harrow from the bottom of the hill to the school-house, made all the difference between easy and difficult travelling, by the change in depth of snow.

8. At the close of the summer, we have then the remnant represented in Fig. 2, on which the snows of the ensuing winter take the form in Fig. 3; and from this greater heap we shall have remaining a greater remnant, which, supposing no wind or other disturbing force modified its form, would appear as at Fig. 4; and, under such necessary modification, together with its own deliquescence, would actually take some such figure as that shown at Fig. 5.

Now, what is there to hinder the continuance of accumulation? If we cover this heap with another layer of winter's snow (Fig. 6), we see at once that the ultimate condition would be, unless somehow prevented, one of enormous mass, superincumbent on the peak—like a colossal haystack, and extending far down its sides below the level of the snow-line.

You are, however, doubtless well aware that no such accumulation as this ever does take place on a mountain-top.

9. So far from it, the eternal snows do not so much as fill the basins between mountain-tops; but, even in these hollows, form depressed sheets at the bottom of them. The difference between the actual aspect of the Alps, and that which they would present if no arrest of the increasing accumulation on them took place, may be shown before you with the greatest ease; and in doing so I have, in all humility, to correct a grave error of my own, which strangely enough, has remained undetected, or at least unaccused, in spite of all the animosity provoked by my earlier writings.

10. When I wrote the first volume of 'Modern Painters,' scarcely any single fact was rightly known by anybody, about either the snow or ice of the Alps. Chiefly the snows had been neglected: very few eyes had ever seen the higher snows near; no foot had trodden the greater number of Alpine summits; and I had to glean what I needed for my pictorial purposes as best I could,—and my best in this case was a blunder. The thing that struck me most, when I saw the Alps myself, was the enormous accumulation of snow on them; and the way it clung to their steep sides. Well, I said to myself, 'of course it must be as thick as it can stand; because, as there is an excess which doesn't melt, it would go on building itself up like the Tower of Babel, unless it tumbled off. There must be always, at the end of winter, as much snow on every high summit as it can carry.'

There *must*, I said. That is the mathematical method

of science as opposed to the artistic. Thinking of a thing, and demonstrating,—instead of looking at it. Very fine, and very sure, if you happen to have before you all the elements of thought; but always very dangerously inferior to the unpretending method of sight—for people who have eyes, and can use them. If I had only *looked* at the snow carefully, I should have seen that it wasn't anywhere as thick as it could stand or lie—or, at least, as a hard substance, though deposited in powder, could stand. And then I should have asked myself, with legitimate rationalism, why it didn't; and if I had but asked——Well, it's no matter what perhaps might have happened if I had. I never did.

11. Let me now show you, practically, how great the error was. Here is a little model of the upper summits of the Bernese range. I shake over them as much flour as they will carry; now I brush it out of the valleys, to represent the melting. Then you see what is left stands in these domes and ridges, representing a mass of snow about six miles deep. That is what the range would be like, however, if the snow stood up as the flour does; and snow is at least, you will admit, as adhesive as flour.

12. But, you will say, the scale is so different, you can't reason from the thing on that scale. A most true objection. You cannot; and therefore I beg you, in like manner, not to suppose that Professor Tyndall's experiments on "a straight prism of ice, four inches long, an inch wide,

and a little more than an inch in depth,"* are conclusive as to the modes of glacier motion.

In what respect then, we have to ask, would the difference in scale modify the result of the experiment made here on the table, supposing this model was the Jungfrau itself, and the flour supplied by a Cyclopean miller and his men?

13. In the first place, the lower beds of a mass six miles deep would be much consolidated by pressure. But would they be *only* consolidated? Would they be in nowise squeezed out at the sides?

The answer depends of course on the nature of flour, and on its conditions of dryness. And you must feel in a moment that, to know what an Alpine range would look like, heaped with any substance whatever, as high as the substance would stand—you must first ascertain how high the given substance *will* stand—on level ground. You might perhaps heap your Alp high with wheat,—not so high with sand,—nothing like so high with dough; and a very thin coating indeed would be the utmost possible result of any quantity whatever of showers of manna, if it had the consistence, as well as the taste, of wafers made with honey.

14. It is evident, then, that our first of inquiries bearing on the matter before us, must be, How high will snow stand on level ground, in a block or column? Suppose

* 'Glaciers of the Alps,' p. 348.

you were to plank in a square space, securely—twenty feet high—thirty—fifty; and to fill it with dry snow. How high could you get your pillar to stand, when you took away the wooden walls? and when you reached your limit, or approached it, what would happen?

Three more questions instantly propose themselves; namely, What happens to snow under given pressure? will it under some degrees of pressure change into anything else than snow? and what length of time will it take to effect the change?

Hitherto, we have spoken of snow as dry only, and therefore as solid substance, permanent in quantity and quality. You know that it very often is not dry; and that, on the Alps, in vast masses, it is throughout great part of the year thawing, and therefore diminishing in quantity.

It matters not the least, to our general inquiry, how much of it is wet, or thawing, or at what times. I merely at present have to introduce these two conditions as elements in the business. It is not dry snow always, but often soppy snow—snow and water,—that you have to squeeze. And it is not freezing snow always, but very often thawing snow,—diminishing therefore in bulk every instant,—that you have to squeeze.

It does not matter, I repeat, to our immediate purpose, when, or how far, these other conditions enter our ground; but it is best, I think, to put the dots on the i's as we go along. You have heard it stated, hinted, suggested, implied, or whatever else you like to call it, again and again, by the modern school of glacialists, that the discoveries of James Forbes were anticipated by Rendu.

15. I have myself more respect for Rendu than any modern glacialist has. He was a man of de Saussure's temper, and of more than de Saussure's intelligence; and if he hadn't had the misfortune to be a bishop, would very certainly have left James Forbes's work a great deal more than cut out for him;—stitched—and pretty tightly—in most of the seams. But he was a bishop; and could only examine the glaciers to an episcopic extent; and guess, the best he could, after that. His guesses are nearly always splendid; but he must needs sometimes reason as well as guess; and he reasons himself with beautiful plausibility, ingenuity, and learning, up to the conclusion—which he announces as positive—that it always freezes on the Alps, even in summer. James Forbes was the first who ascertained the fallacy of this episcopal position; and who announced—to our no small astonishment—that it always thawed on the Alps, even in winter.

16. Not superficially of course, nor in all places. But internally, and in a great many places. And you will find it is an ascertained fact—the first great one of which we owe the discovery to him—that all the year round, you must reason on the masses of aqueous deposit on the Alps as, practically, in a state of squash. Not freezing ice or snow, nor dry ice or snow, but in many places saturated with,—everywhere affected by,—moisture; and always

subject, in enormous masses, to the conditions of change which affect ice or snow at the freezing-point, and not below it. Even James Forbes himself scarcely, I think, felt enough the importance of this element of his own discoveries, in all calculations of glacier motion. He sometimes speaks of his glacier a little too simply as if it were a stream of *undiminishing* substance, as of treacle or tar, moving under the action of gravity only; and scarcely enough recognizes the influence of the subsiding languor of its fainting mass, as a constant source of motion; though nothing can be more accurate than his actual account of its results on the surface of the Mer de Glace, in his fourth letter to Professor Jameson.

17. Let me drive the notion well home in your own minds, therefore, before going farther. You may permanently secure it, by an experiment easily made by each one of you for yourselves this evening, and that also on the minute and easily tenable scale which is so approved at the Royal Institution; for in this particular case the material conditions may indeed all be represented in very small compass. Pour a little hot water on a lump of sugar in your teaspoon. You will immediately see the mass thaw, and subside by a series of, in miniature, magnificent and appalling catastrophes, into a miniature glacier, which you can pour over the edge of your teaspoon into your saucer; and if you will then add a little of the brown sugar of our modern commerce—of a slightly sandy character,—you may watch the rate of the flinty erosion upon the soft silver

of the teaspoon at your ease, and with Professor Ramsay's help, calculate the period of time necessary to wear a hole through the bottom of it.

I think it would be only tiresome to you if I carried the inquiry farther by progressive analysis. You will, I believe, permit, or even wish me, rather to state summarily what the facts are:—their proof, and the process of their discovery, you will find incontrovertibly and finally given in this volume, classical, and immortal in scientific literature—which, twenty-five years ago, my good master Dr. Buckland ordered me, in his lecture-room at the Ashmolean, to get,—as closing all question respecting the nature and cause of glacier movement,—James Forbes's 'Travels in the Alps.'

18. The entire mass of snow and glacier, (the one passing gradually and by infinite modes of transition into the other, over the whole surface of the Alps,) is one great accumulation of ice-cream, poured upon the tops, and *flowing* to the bottoms, of the mountains, under precisely the same special condition of gravity and coherence as the melted sugar poured on the top of a bride-cake; but on a scale which induces forms and accidents of course peculiar to frozen water, as distinguished from frozen syrup, and to the scale of Mont Blanc and the Jungfrau, as compared to that of a bride-cake. Instead of an inch thick, the ice-cream of the Alps will stand two hundred feet thick,—no thicker, anywhere, if it can run off; but will lie in the hollows like lakes, and clot and cling about the less abrupt slopes in festooned wreaths of rich mass and

sweeping flow, breaking away, where the steepness becomes intolerable, into crisp precipices and glittering cliffs.

19. Yet never for an instant motionless—never for an instant without internal change, through all the gigantic mass, of the relations to each other of every crystal grain. That one which you break now from its wave-edge, and which melts in your hand, has had no rest, day nor night, since it faltered down from heaven when you were a babe at the breast; and the white cloud that scarcely veils yonder summit—seven-colored in the morning sunshine—has strewed it with pearly hoar-frost, which will be on this spot, trodden by the feet of others, in the day when you also will be trodden under feet of men, in your grave.

20. Of the infinite subtlety, the exquisite constancy of this fluid motion, it is nearly impossible to form an idea in the least distinct. We hear that the ice advances two feet in the day; and wonder how such a thing can be possible, unless the mass crushed and ground down everything before it. But think a little. Two feet in the day is a foot in twelve hours,—only an inch in an hour, (or say a little more in the daytime, as less in the night,)—and that is maximum motion in mid-glacier. If your Geneva watch is an inch across, it is three inches round, and the minute-hand of it moves three times faster than the fastest ice. Fancy the motion of that hand so slow that it must take three hours to get round the little dial. Between the shores of this vast gulf of hills, the long wave of hastening ice only keeps pace with that lingering

arrow, in its central crest; and that invisible motion fades away upwards through forty years of slackening stream, to the pure light of dawn on yonder stainless summit, on which this morning's snow lies—motionless.

21. And yet, slow as it is, this infinitesimal rate of current is enough to drain the vastest gorges of the Alps of their snow, as clearly as the sluice of a canal-gate empties a lock. The mountain basin included between the Aiguille Verte, the Grandes Jorasses, and the Mont Blanc, has an area of about thirty square miles, and only one outlet, little more than a quarter of a mile wide: yet, through this the contents of the entire basin are drained into the valley of Chamounix with perfect steadiness, and cannot possibly fill the basin beyond a certain constant height above the point of overflow.

Overflow, I say, deliberately; distinguishing always the motion of this true fluid from that of the sand in an hourglass, or of stones slipping in a heap of shale. But that the nature of this distinction may be entirely conceived by you, I must ask you to pause with some attention at this word, to 'flow,'—which attention may perhaps be more prudently asked in a separate chapter.

# CHAPTER IV.

## LABITUR, ET LABETUR.

(*Lecture given at London Institution, continued, with added Illustrations.*)

1. Of course—we all know what flowing means. Well, it is to be hoped so; but I'm not sure. Let us see. The sand of the hour-glass,—do you call the motion of that flowing?

No. It is only a consistent and measured fall of many unattached particles.

Or do you call the entrance of a gas through an aperture, out of a full vessel into an empty one, flowing?

No. That is expansion—not flux.

Or the draught through the keyhole? No—is your answer, still. Let us take instance in water itself. The *spring* of a fountain, or of a sea breaker into spray. You don't call that flowing?

No.

Nor the *fall* of a fountain, or of rain?

No.

Well, the *rising* of a breaker,—the current of water in the hollow shell of it,—is *that* flowing? No. After it has broken—rushing up over the shingle, or impatiently advancing on the sand? You begin to pause in your negative.

Drooping back from the shingle then, or ebbing from the sand? Yes; flowing, in some places, certainly, now.

You see how strict and distinct the idea is in our minds. Will you accept—I think you may,—this definition of it? Flowing is "the motion of liquid or viscous matter over solid matter, under the action of gravity, without any other impelling force."

2. Will you accuse me, in pressing this definition on you, of wasting time in mere philological nicety? Permit me, in the capacity which even the newspapers allow to me,—that of a teacher of expression,—to answer you, as often before now, that philological nicety is philosophical nicety. See the importance of it here. I said a glacier flowed. But it remains a question whether it does not also *spring*,—whether it can rise as a fountain, no less than descend as a stream.

For, broadly, there are two methods in which either a stream or glacier moves.

The first, by withdrawing a part of its mass in front, the vacancy left by which, another part supplies from behind.

That is the method of a continuous stream,—perpetual deduction,* by what precedes, of what follows.

* "Ex quo illa admirabilis a majoribus aquæ facta deductio est."—Cic. de Div., 1, 44.

The second method of motion is when the mass that is behind, presses, or is poured in upon, the masses before. That is the way in which a cataract falls into a pool, or a fountain into a basin.

Now, in the first case, you have catenary curves, or else curves of traction, going down the stream. In the second case, you have irregularly concentric curves, and ripples of impulse and compression, succeeding each other round the pool.

3. Now the Mer de Glace is deduced down its narrow channel, like a river; and the Glacier des Bossons is deduced down its steep ravine; and both were once injected into a pool of ice in the valley below, as the Glacier of the Rhone is still. Whereupon, observe, if a stream falls into a basin—level-lipped all round—you know when it runs over it must be pushed over—lifted over. But if ice is thrown into a heap in a plain, you can't tell, without the closest observation, how violently it is pushed from behind, or how softly it is diffusing itself in front; and I had never set my eyes or wits to ascertain where compression in the mass ceased, and diffusion began, because I thought Forbes had done everything that had to be done in the matter. But in going over his work again I find he has left just one thing to be still explained; and that one chances to be left to me to show you this evening, because, by a singular and splendid Nemesis, in the obstinate rejection of Forbes's former conclusively simple experiments, and in the endeavour to substitute others of his own, Pro-

fessor Tyndall has confused himself to the extreme point of not distinguishing these two conditions of deductive and impulsive flux. His incapacity of drawing, and ignorance of perspective, prevented him from constructing his diagrams either clearly enough to show him his own mistakes, or prettily enough to direct the attention of his friends to them;—and they luckily remain to us, in their absurd immortality.

4. Forbes poured viscous substance in layers down a trough; let the stream harden; cut it into as many sections as were required; and showed, in permanence, the actual conditions of such viscous motion. Eager to efface the memory of these conclusive experiments, Professor Tyndall ('Glaciers of the Alps,' page 383) substituted this literally 'superficial' one of his own. He stamped circles on the top of a viscous current; found, as it flowed, that they were drawn into ovals; but had not wit to consider, or sense to see, whether the area of the circle was enlarged or diminished—or neither—during its change in shape. He jumped, like the rawest schoolboy, to the conclusion that a circle, becoming an oval, must necessarily be compressed! You don't compress a globe of glass when you blow it into a soda-water bottle, do you?

5. But to reduce Professor Tyndall's problem into terms. Let A F, Fig. 3 (page 54), be the side of a stream of any substance whatever, and *a f* the middle of it; and let the particles at the middle move twice as fast as

Fig. 3.

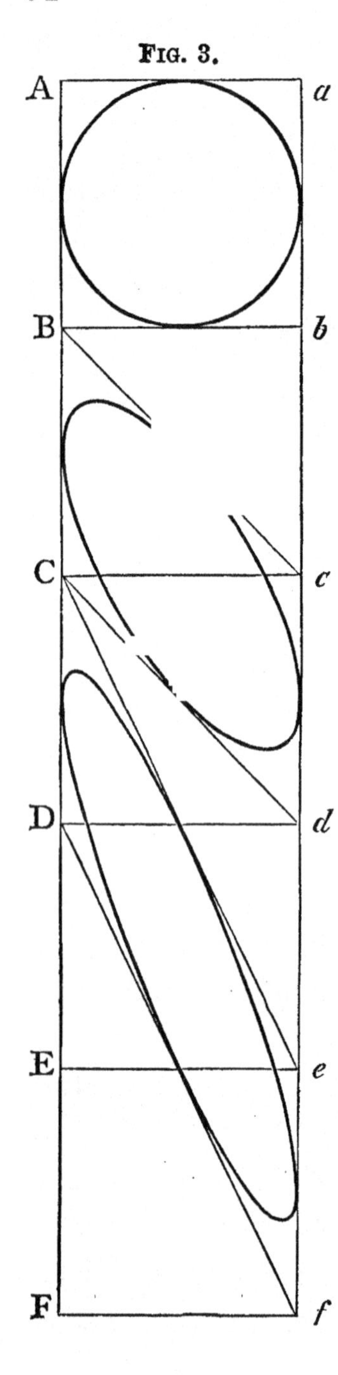


the particles at the sides. Now we cannot study all the phenomena of fluid motion in one diagram, nor any one phenomenon of fluid motion but by progressive diagrams; and this first one only shows the changes of form which would take place in a substance which moved with *uniform* increase of rapidity from side to centre. No fluid substance *would* so move; but you can only trace the geometrical facts step by step, from uniform increase to accelerated increase. Let the increase of rapidity, therefore, first be supposed uniform. Then, while the point A moves to B, the point *a* moves to *c*, and any points once intermediate in a right line between A and *a*, will now be intermediate in a right line between B and *c*, and their places determinable by verticals from each to each.

I need not be tedious in farther describing the figure. Suppose A *b* a square mile of the substance, and the origin of motion on the line A *a*. Then when the point A has arrived at B, the point B has arrived at C, the point *a* at *c*, and the point *b* at *d*, and the mile square, A *b*, has become the mile rhombic, B *d*, of the same area; and if there were a circle drawn in the square A *b*, it will become the fat ellipse in B *d*, and thin ellipse in C *f*, successively.

6. Compressed, thinks Professor Tyndall, one way, and stretched the other!

But the Professor has never so much as understood what 'stretching' means. He thinks that ice won't stretch! Does he suppose treacle, or oil, *will?* The brilliant natural philosopher has actually, all through his two books on glaciers, confused viscosity with elasticity! You can *stretch* a piece of Indian-rubber, but you can only *diffuse* treacle, or oil, or water.

"But you can draw these out into a narrow stream, whereas you cannot pull the ice?"

No; neither can you pull water, can you? In compressing any substance, you can apply any force you like; but in extending it, you can only apply force less than that with which its particles cohere. You can pull honey into a thin string, when it comes out of the comb; let it be candied, and you can't pull it into a thin string. Does that make it less a viscous substance? You can't stretch mortar either. It cracks even in the hod, as it is heaped. Is it, therefore, less fluent or manageable in the mass?

7. Whereas the curious fact of the matter is, that, in

precise contrariety to Mr. Tyndall's idea, ice, (glacier ice, that is to say,) *will* stretch; and that treacle or water won't! and that's just the plague of dealing with the whole glacier question—that the incomprehensible, untenable, indescribable ice will both squeeze and open; and is slipping through your fingers all the time besides, by melting away. You can't deal with it as a simple fluid; and still less as a simple solid. And instead of having less power to accommodate itself to the irregularities of its bed than water, it has much more;—a great deal more of it will subside into a deep place, and ever so much of it melt in passing over a shallow one; and the centre, at whatever rate it moves, will supply itself by the exhaustion of the sides, instead of raging round, like a stream in back-water.

8. However, somehow, I must contrive to deal at least with the sure fact that the velocity of it is progressively greater from the sides to the centre, and from the bottom to the surface.

Now it is the last of these progressive increments which is of chief importance to my present purpose.

For my own conviction on the matter;—mind, not *theory*, for a man can always avoid constructing theories, but cannot possibly help his convictions, and may sometimes feel it right to state them,—my own conviction is that the ice, when it is of any considerable depth, no more moves over the bottom than the lower particles of a running stream of honey or treacle move over a plate but that, in entire rest at the bottom, except so far as it is moved by dissolution, it increases in velocity to the surface in a curve of the nature of a parabola, or a logarithmic curve, capable of being infinitely prolonged, on the supposition of the depth of the ice increasing to infinity.

9. But it is now my fixed principle not to care what I think, when a fact can be ascertained by looking, or measuring. So, not having any observations of my own on this matter, I seek what help may be had elsewhere; and find in the eleventh chapter of Professor Tyndall's 'Glaciers of the Alps,' two most valuable observations, made under circumstances of considerable danger, calmly encountered by the author, and grumblingly by his guide,— danger consisting in the exposure to a somewhat close and well-supported fire of round and grape from the glacier of the Géant, which objected to having its velocity measured. But I find the relation of these adventures so much distract me from the matter in hand, that I must digress briefly into some notice of the general literary structure of this remarkable book.

10. Professor Tyndall never fails to observe with complacency, and to describe to his approving readers, how unclouded the luminous harmonies of his reason, imagination, and fancy remained, under conditions which, he rightly concludes, would have been disagreeably exciting, or even distinctly disturbing, to less courageous persons. And indeed I confess, for my own part, that my success-

fullest observations have always been made while lying all my length on the softest grass I could find; and after assuring myself with extreme caution that if I chanced to go to sleep, (which in the process of very profound observations I usually do, at least of an afternoon,) I am in no conceivable peril beyond that of an ant-bite. Nevertheless, the heroic Professor does not, it seems to me, sufficiently recognize the universality of the power of English, French, German, and Italian gentlemen to retain their mental faculties under circumstances even of more serious danger than the crumbling of a glacier moraine; and to think with quickness and precision, when the chances of death preponderate considerably, or even conclusively, over those of life. Nor does Professor Tyndall seem to have observed that the gentlemen possessing this very admirable power in any high degree, do not usually think their own emotions, or absence of emotions, proper subjects of printed history, and public demonstration.

11. Nevertheless, when a national philosopher, under showers of granite grape, places a stake and auger against his heart, buttons his coat upon them, and cuts himself an oblique staircase up a wall of ice, nearly vertical, to a height of forty feet from the bottom; and there, unbuttoning his coat, pierces the ice with his auger, drives in his stake, and descends without injury, though during the whole operation his guide "growls audibly," we are bound to admit his claim to a scientific Victoria Cross—or at least crosslet,— and even his right to walk about in our

London drawing-rooms in a gracefully cruciferous costume; while I have no doubt also that many of his friends will be interested in such metaphysical particulars and examples of serene mental analysis as he may choose to give them in the course of his autobiography. But the Professor ought more clearly to understand that scientific writing is one thing, and pleasant autobiography another; and though an officer may not be able to give an account of a battle without involving some statement of his personal share in it, a scientific observer might with entire ease, and much convenience to the public, have published 'The Glaciers of the Alps' in two coincident, but not coalescing, branches—like the glaciers of the Giant and Léchaud; and that out of the present inch and a half thickness of the volume, an inch and a quarter might at once have been dedicated to the Giant glacier of the autobiography, and the remaining quarter of an inch to the minor current of scientific observation, which, like the Glacier de Léchaud, appears to be characterized by "the comparative shallowness of the upper portion," * and by its final reduction to "a driblet measuring about one-tenth of its former transverse dimensions."

12. It is true that the book is already divided into two portions,—the one described as "chiefly narrative," and the other as "chiefly scientific." The chiefly narrative portion is, indeed, full of very interesting matter fully jus-

* 'Glaciers of the Alps,' p. 288.

tifying its title; as, for instance, "We tumbled so often in the soft snow, and our clothes and boots were so full of it, that we thought we might as well try the sitting posture in sliding down. We did so, and descended with extraordinary velocity" (p. 116). Or again: "We had some tea, which had been made at the Montanvert, and carried up to the Grand Mulets in a bottle. My memory of that tea is not pleasant" (p. 73). Or in higher strains of scientific wit and pathos: "As I looked at the objects which had now become so familiar to me, I felt that, though not viscous, the ice did not lack the quality of adhesiveness, and I felt a little sad at the prospect of bidding it so soon farewell."

13. But the merely romantic readers of this section, rich though it be in sentiment and adventure, will find themselves every now and then arrested by pools, as it were, of almost impassable scientific depth—such as the description of a rock "evidently to be regarded as an assemblage of magnets, or as a single magnet full of consequent points" (p. 140). While, on the other hand, when in the course of my own work, finding myself pressed for time, and eager to collect every scrap of ascertained data accessible to me, I turn hopefully to the eleventh chapter of the "chiefly scientific" section of the volume, I think it hard upon me that I must read through three pages of narrative describing the Professor's dangers and address, before I can get at the two observations which are the sum of the scientific contents of the chapter, yet to the first of

which "unfortunately some uncertainty attached itself," and the second of which is wanting in precisely the two points which would have made it serviceable. First, it does not give the rate of velocity at the base, but five feet above the base; and, secondly, it gives only three measurements of motion. Had it given four, we could have drawn the curve: but we can draw any curve we like through three points.

14. I will try the three points, however, with the most probable curve; but this being a tedious business, will reserve it for a separate chapter, which readers may skip if they choose: and insert, for the better satisfaction of any who may have been left too doubtful by the abrupt close of my second chapter, this postscript, written the other day after watching the streamlets on the outlying fells of Shap.

15. Think what would be the real result, if any stream among our British hills at this moment *were* cutting its bed deeper.

In order to do so, it must of course annually be able to remove the entire zone of débris moved down to its bed from the hills on each side of it—and somewhat more.

Take any Yorkshire or Highland stream you happen to know, for example; and think what quantity of débris must be annually moved, on the hill surfaces which feed its waters. Remember that a lamb cannot skip on their slopes, but it stirs with its hoofs some stone or grain of dust which will more or less roll or move downwards.

That no shower of rain can fall—no wreath of snow melt, without moving some quantity of dust downwards. And that no frost can break up, without materially loosening some vast ledges of crag, and innumerable minor ones; nor without causing the fall of others as vast, or as innumerable. Make now some effort to conceive the quantity of rock and dust moved annually, lower, past any given level traced on the flanks of any considerable mountain stream, over the area it drains—say, for example, in the basin of the Ken above Kendal, or of the Wharfe above Bolton Abbey.

16. Then, if either of those streams were cutting their beds deeper,—that quantity of rock, and something more, must be annually carried down by their force, past Kendal bridge, and Bolton stepping-stones. Which you will find would occasion phenomena very astonishing indeed to the good people of Kendal and Wharfedale.

17. "But it need not be carried down past the stepping-stones," you say—"it may be deposited somewhere above." Yes, that is precisely so;—and wherever it *is* deposited, the bed of the stream, or of some tributary streamlet, is being raised. Nobody notices the raising of it;—another stone or two among the wide shingle—a tongue of sand an inch or two broader at the burnside—who can notice that? Four or five years pass;—a flood comes;—and Farmer So-and-So's field is covered with slimy ruin. And Farmer So-and-So's field is an inch higher than it was, for evermore—but who notices that? The shingly stream has

gone back into its bed: here and there a whiter stone or two gleams among its pebbles, but next year the water stain has darkened them like the rest, and the bed is just as far below the level of the field as it was. And your careless geologist says, 'what a powerful stream it is, and how deeply it is cutting its bed through the glen!'

18. Now, carry out this principle for existing glaciers. If the glaciers of Chamouni were cutting their beds deeper, either the annual line of débris of the Mont Blanc range on its north side must be annually carried down past the Pont Pelissier; or the valley of Chamouni must be in process of filling up, while the ravines at its sides are being cut down deeper. Will any geologist, supporting the modern glacial theories, venture to send me, for the next number of Deucalion, his idea, on this latter, by him inevitable, hypothesis, of the profile of the bottom of the Glacier des Bossons, a thousand years ago; and a thousand years hence?

# CHAPTER V.

## THE VALLEY OF CLUSE.

1. WHAT strength of faith men have in each other; and how impossible it is for them to be independent in thought, however hard they try! Not that they ever ought to be; but they should know, better than they do, the incumbrance that the false notions of others are to them.

Touching this matter of glacial grinding action; you will find every recent writer taking up, without so much as a thought of questioning it, the notion adopted at first careless sight of a glacier stream by some dull predecessor of all practical investigation—that the milky colour of it is all produced by dust ground off the rocks at the bottom. And it never seems to occur to any one of the Alpine Club men, who are boasting perpetually of their dangers from falling stones; nor even to professors impeded in their most important observations by steady fire of granite grape, that falling stones may probably knock their edges off when they strike; and that moving banks and fields of moraine, leagues long, and leagues square, of which every stone is shifted a foot forward every day on a surface melting beneath them, must in such shifting be liable to attrition enough to produce considerably more dust, and that of the finest kind, than any glacier stream carries down with it—not to speak of processes of decomposition accelerated, on all surfaces liable to them, by alternate action of frost and fierce sunshine.

2. But I have not, as yet, seen any attempts to determine even the first data on which the question of attrition must be dealt with. I put it, in simplicity, at the close of last chapter. But, in its full extent, the inquiry ought not to be made merely of the bed of the Glacier des Bossons; but of the bed of the Arve, from the Col de Balme to Geneva; in which the really important points for study are the action of its waters at Pont Pelissier;—at the falls below Servoz;—at the portal of Cluse;—and at the northern end of the slope of the Salève.

3. For these four points are the places where, if at all, sculptural action is really going on upon its bed: at those points, if at all, the power of the Second Æra, the æra of sculpture, is still prolonged into this human day of ours. As also it is at the rapids and falls of all swiftly descending rivers. The one vulgar and vast deception of Niagara has blinded the entire race of modern geologists to the primal truth of mountain form, namely, that the rapids and cascades of their streams indicate, not points to which the falls have receded, but places where the remains of once colossal cataracts still exist, at the places eternally (in human experience) appointed for the formation of such cataracts, by the form and hardness of the local rocks. The rapids of the Amazon, the Nile, and the Rhine, obey

precisely the same law as the little Wharfe at its Strid, or as the narrow 'rivus aquæ' which, under a bank of strawberries in my own tiny garden, has given me perpetual trouble to clear its channel of the stones brought down in flood, while, just above, its place of picturesque cascade, is determined for it by a harder bed of Coniston flags, and the little pool, below that cascade, never encumbered with stones at all.

4. Now the bed of the Arve, from the crest of the Col de Balme to Geneva, has a fall of about 5,000 feet; and if any young Oxford member of the Alpine Club is minded to do a piece of work this vacation, which in his old age, when he comes to take stock of himself, and edit the fragments of himself, as I am now sorrowfully doing, he will be glad to have done, (even though he risked neither his own nor any one else's life to do it,) let him survey that bed accurately, and give a profile of it, with the places and natures of emergent rocks, and the ascertainable depths and dates of alluvium cut through, or in course of deposition.

5. After doing this piece of work carefully, he will probably find some valuable ideas in his head concerning the proportion of the existing stream of the Arve to that which once flowed from the glacier which deposited the moraine of Les Tines; and again, of that torrent to the infinitely vaster one of the glacier that deposited the great moraine of St. Gervais; and finally of both, to the cliffs of Cluse, which have despised and resisted them. And ideas

which, after good practical work, he finds in his head, are likely to be good for something: but he must not seek for them; all thoughts worth having come like sunshine, whether we will or no: the thoughts not worth having, are the little lucifer matches we strike ourselves.

6. And I hasten the publication of this number of Deucalion, to advise any reader who cares for the dreary counsel of an old-fashioned Alpine traveller, to see the valley of Cluse this autumn, if he may, rather than any other scene among the Alps;—for if not already destroyed, it must be so, in a few months more, by the railway which is to be constructed through it, for the transport of European human diluvium. The following note of my last walk there, written for my autumn lectures, may be worth preserving among the shingle of my scattered work.

7. I had been, for six months in Italy, never for a single moment quit of liability to interruption of thought. By day or night, whenever I was awake, in the streets of every city, there were entirely monstrous and inhuman noises in perpetual recurrence. The violent rattle of carriages, driven habitually in brutal and senseless haste, or creaking and thundering under loads too great for their cattle, urged on by perpetual roars and shouts: wild bellowing and howling of obscene wretches far into the night: clashing of church bells, in the morning, dashed into reckless discord, from twenty towers at once, as if rung by devils to defy and destroy the quiet of God's sky,

and mock the laws of His harmony: filthy, stridulous shrieks and squeaks, reaching for miles into the quiet air, from the railroad stations at every gate: and the vociferation, endless, and frantic, of a passing populace whose every word was in mean passion, or in unclean jest. Living in the midst of this, and of vulgar sights more borrible than the sounds, for six months, I found myself—suddenly, as in a dream—walking again alone through the valley of Cluse, unchanged since I knew it first, when I was a boy of fifteen, quite forty years ago;—and in perfect quiet, and with the priceless completion of quiet, that I was without fear of any outcry or base disturbance of it.

8. But presently, as I walked, the calm was deepened, instead of interrupted, by a murmur—first low, as of bees, and then rising into distinct harmonious chime of deep bells, ringing in true cadences—but I could not tell where. The cliffs on each side of the valley of Cluse vary from 1,500 to above 2,000 feet in height; and, without absolutely echoing the chime, they so accepted, prolonged, and diffused it, that at first I thought it came from a village high up and far away among the hills; then presently it came down to me as if from above the cliff under which I was walking; then I turned about and stood still, wondering; for the whole valley was filled with the sweet sound, entirely without local or conceivable origin: and only after some twenty minutes' walk, the depth of tones, gradually increasing, showed me that they came from the

tower of Maglans in front of me; but when I actually got into the village, the cliffs on the other side so took up the ringing, that I again thought for some moments I was wrong.

Perfectly beautiful, all the while, the sound, and exquisitely varied,—from ancient bells of perfect tone and series, rung with decent and joyful art.

"What are the bells ringing so to-day for,—it is no fête?" I asked of a woman who stood watching at a garden gate.

"For a baptism, sir."

And so I went on, and heard them fading back, and lost among the same bewildering answers of the mountain air.

9. Now that half-hour's walk was to me, and I think would have been to every man of ordinarily well-trained human and Christian feeling—I do not say merely worth the whole six months of my previous journey in Italy;—it was a reward for the endurance and horror of the six months' previous journey; but, as many here may not know what the place itself is like, and may think I am making too much of a little pleasant bell-ringing, I must tell you what the valley of Cluse is in itself.

10. Of 'Cluse,' the closed valley,—not a ravine, but a winding plain, between very great mountains, rising for the most part in cliffs—but cliffs which retire one behind the other above slopes of pasture and forest. (Now as I am writing this passage in a country parsonage—of Cow-

ley, near Uxbridge,—I am first stopped by a railroad whistle two minutes and a half long,* and then by the rumble and grind of a slow train, which prevents me from hearing my own words, or being able to think, so that I must simply wait for ten minutes, till it is past.)

It being past, I can go on. Slopes of pasture and forest, I said, mingled with arable land, in a way which you can only at present see in Savoy; that is to say, you have walnut and fruit trees of great age, mixed with oak, beech, and pine, as they all choose to grow—it seems as if the fruit trees planted themselves as freely as the pines. I imagine this to be the consequence of a cultivation of very ancient date under entirely natural laws; if a plum-tree or a walnut planted itself, it was allowed to grow; if it came in the way of anything or anybody, it would be cut down; but on the whole the trees grew as they liked; and the fields were cultivated round them in such spaces as the rocks left;—ploughed, where the level admitted, with a ploughshare lightly constructed, but so huge that it looks more like the beak of a trireme than a plough, two oxen forcing it to heave aside at least two feet depth of the light earth;—no fences anywhere; winding field walks, or rock paths, from cottage to cottage; these last not of the luxurious or trim Bernese type, nor yet comfortless châlets; but sufficient for orderly and virtuous life: in outer aspect, beautiful exceedingly, just because

* Counted by watch, for I knew by its manner it would last, and measured it.

their steep roofs, white walls, and wandering vines had no pretence to perfectness, but were wild as their hills. All this pastoral country lapped into inlets among the cliffs, vast belts of larch and pine cresting or clouding the higher ranges, whose green meadows change as they rise, into mossy slopes, and fade away at last among the grey ridges of rock that are soonest silvered with autumnal snow.

11. The ten-miles length of this valley, between Cluse and St. Martin's, include more scenes of pastoral beauty and mountain power than all the poets of the world have imagined; and present more decisive and trenchant questions respecting mountain structure than all the philosophers of the world could answer: yet the only object which occupies the mind of the European travelling public, respecting it, is to get through it, if possible, under the hour.

12. I spoke with sorrow, deeper than my words attempted to express, in my first lecture, of the blind rushing of our best youth through the noblest scenery of the Alps, without once glancing at it, that they might amuse, or kill, themselves on their snow. That the claims of all sweet pastoral beauty, of all pious domestic life, for a moment's pause of admiration or sympathy, should be unfelt, in the zest and sparkle of boy's vanity in summer play, may be natural at all times; and inevitable while our youth remain ignorant of art, and defiant of religion; but that, in the present state of science, when every eye

is busied with the fires in the Moon and the shadows in the Sun, no eye should occupy itself with the ravines of its own world, nor with the shadows which the sun casts on the cliffs of them; that the simplest,—I do not say problems, but bare facts, of structure,—should still be unrepresented, and the outmost difficulties of rock history untouched; while dispute, and babble, idler than the chafed pebbles of the wavering beach, clink, jar, and jangle on from year to year in vain,—surely this, in our great University, I am bound to declare to be blameful; and to ask you, with more than an artist's wonder, why this fair valley of Cluse is now closed indeed, and forsaken, "clasped like a missal shut where Paynims pray;" and, with all an honest inquirer's indignation, to challenge—in the presence of our Master of Geology, happily one of its faithful and true teachers,* the Speakers concerning the Earth,—the geologists, not of England only, but of Europe and America,—either to explain to you the structure or sculpture of this † renownedest cliff in all the Alps, under which Tell leaped ashore; or to assign valid reason for the veins in the pebbles which every Scotch lassie wears for her common jewellery.

---

* Mr. Prestwich. I have to acknowledge, with too late and vain gratitude, the kindness and constancy of the assistance given me, on all occasions when I asked it, by his lamented predecessor in the Oxford Professorship of Geology, Mr. Phillips.

† The cliff between Fluelen and Brunnen, on the lake of Uri, of which Turner's drawing was exhibited at this lecture.

## CHAPTER VI.

### OF BUTTER AND HONEY.

1. The last chapter, being properly only a continuation of the postscript to the fourth, has delayed me so long from my question as to ice-curves, that I cannot get room for the needful diagrams and text in this number; which is perhaps fortunate, for I believe it will be better first to explain to the reader more fully why the ascertainment of this curve of vertical motion is so desirable.

To which explanation, very clear definition of some carelessly used terms will be essential.

2. The extremely scientific Professor Tyndall always uses the terms Plastic, and Viscous, as if they were synonymous. But they express entirely different conditions of matter The first is the term proper to be used of the state of butter, on which you can stamp whatever you choose; and the stamp will stay; the second expresses that of honey, on which you can indeed stamp what you choose; but the stamp melts away forthwith.

And of viscosity itself there are two distinct varieties—one glutinous, or gelatinous, like that of treacle or tapioca soup; and the other simply adhesive, like that of mercury or melted lead.

And of both plasticity and viscosity there are infinitely various degrees in different substances, from the perfect and absolute plasticity of gold, to the fragile, and imperfect, but to man more precious than any quality of gold, plasticity of clay, and, most precious of all, the blunt and dull plasticity of dough; and again, from the vigorous and binding viscosity of stiff glue, to the softening viscosity of oil, and tender viscosity of old wine. I am obliged therefore to ask my readers to learn, and observe very carefully in our future work, these following definitions.

*Plastic.*—Capable of change of form under external force, without any loss of continuity of substance; and of *retaining afterwards the form imposed on it.*

Gold is the most perfectly plastic substance we commonly know; clay, butter, etc., being more coarsely and ruggedly plastic, and only in certain consistencies or at certain temperatures.

*Viscous.*—Capable of change of form under external force, *but not of retaining the form imposed;* being languidly obedient to the force of gravity, and necessarily declining to the lowest possible level,—as lava, treacle, or honey.

*Ductile.*—Capable of being extended by traction without loss of continuity of substance. Gold is both plastic and ductile; but clay, plastic only, not ductile; while most melted metals are ductile only, but not plastic.

*Malleable.*—Plastic only under considerable force.

3. We must never let any of these words entangle, as necessary, the idea belonging to another.

A plastic substance is not necessarily ductile, though gold is both; a viscous substance is not necessarily ductile, though treacle is both; and the quality of elasticity, though practically inconsistent with the character either of a plastic body, or a viscous one, may enter both the one and the other as a gradually superadded or interferent condition, in certain states of congelation; as in indian-rubber, glass, sealing-wax, asphalt, or basalt.

I think the number of substances I have named in this last sentence, and the number of entirely different states which in an instant will suggest themselves to you, as characteristic of each, at, and above, its freezing or solidifying point, may show at once how careful we should be in defining the notion attached to the words we use; and how inadequate, without specific limitation and qualification, *any* word must be, to express all the qualities of any given substance.

4. But, above all substances that can be proposed for definition of quality, glacier ice is the most defeating. For it is practically plastic; but *actually* viscous;—and that to the full extent. You can beat or hammer it, like gold; and it will stay in the form you have beaten it into, for a time;—and so long a time, that, on all instant occasions of plasticity, it is practically plastic. But only have patience to wait long enough, and it will run down out of the form you have stamped on it, as honey does

so that, actually and inherently, it is viscous, and not plastic.

5. Here then, at last, I have got Forbes's discovery and assertion put into accurately intelligible terms;—very incredible terms, I doubt not, to most readers.

There is not the smallest hurry, however, needful in believing them: only let us understand clearly what it is we either believe or deny; and in the meantime, return to our progressive conditions of snow on the simplest supposable terms, as shown in my first plate.

6. On a conical mountain, such as that represented in Fig. 6, we are embarrassed by having to calculate the subtraction by avalanche down the slopes. Let us therefore take rather, for examination, a place where the snow can lie quiet.

Let Fig. 7, Plate I., represent a hollow in rocks at the summit of a mountain above the line of perpetual snow, the lowest watershed being at the level indicated by the dotted line. Then the snow, once fallen in this hollow, can't get out again; but a little of it is taken away every year, partly by the heat of the ground below, partly by surface sunshine and evaporation, partly by filtration of water from above, while it is also saturated with water in thaw-time, up to the level of watershed. Consequently it must subside every year in the middle; and, as the mass remains unchanged, the same quantity must be added every year at the top,—the excess being always, of course, blown away, or dropped off, or thawed above, in the year it falls.

7. Hence the entire mass will be composed, at any given time, of a series of beds somewhat in the arrangement given in Fig. 8; more remaining of each year's snow in proportion to its youth, and very little indeed of the lowest and oldest bed.

It *must* subside, I say, every year;—but how much is involved, of new condition, in saying this! Take the question in the simplest possible terms; and let Fig. 9 represent a cup or crater full of snow, level in its surface at the end of winter. During the summer, there will be large superficial melting; considerable lateral melting by reverberation from rock, and lateral drainage; bottom melting from ground heat, not more than a quarter of an inch,—(Forbes's Travels, page 364,)—a quantity which we may practically ignore. Thus the mass, supposing the substance of it immovable in position, would be reduced by *superficial* melting during the year to the form approximately traced by the dotted line within it, in Fig. 9.

8. But how of the *interior* melting? Every interstice and fissure in the snow, during summer, is filled either with warm air, or warm water in circulation through it, and every separate surface of crystal is undergoing its own degree of diminution. And a constant change in the conditions of equilibrium results on every particle of the mass; and a constant subsidence takes place, involving an entirely different relative position of every portion of it at the end of the year.

9. But I cannot, under any simple geometrical figure,

give an approximation to the resultant directions of change in form; because the density of the snow must be in some degree proportioned to the depth, and the melting less, in proportion to the density.

Only at all events, towards the close of the year, the mass enclosed by the dotted line in Fig. 9 will have sunk into some accommodation of itself to the hollow bottom of the crater, as represented by the continuous line in Fig. 10. And, over that, the next winter will again heap the snow to the cup-brim, to be reduced in the following summer; but now through two different states of consistence, to the bulk limited by the dotted line in Fig. 10.

10. In a sequence of six years, therefore, we shall have a series of beds approximately such as in Fig. 11;—ap proximately observe, I say always, being myself wholly unable to deal with the complexities of the question, and only giving the diagram for simplest basis of future investigation, by the first man of mathematical knowledge and praetical common sense, who will leave off labouring for the contradiction of his neighbours, and apply himself to the hitherto despised toil of the ascertainment of facts. And when he has determined what the positions of the strata will be in a perfectly uniform cup, such as that of which the half is represented in perspective in Fig. 12, let him next inquire what would have happened to the mass, if, instead of being deposited in a cup enclosed, on all sides, it had been deposited in an amphitheatre open on one, as in the section shown in Fig. 12. For that is indeed the first radical problem to be determined respecting glacier motion.

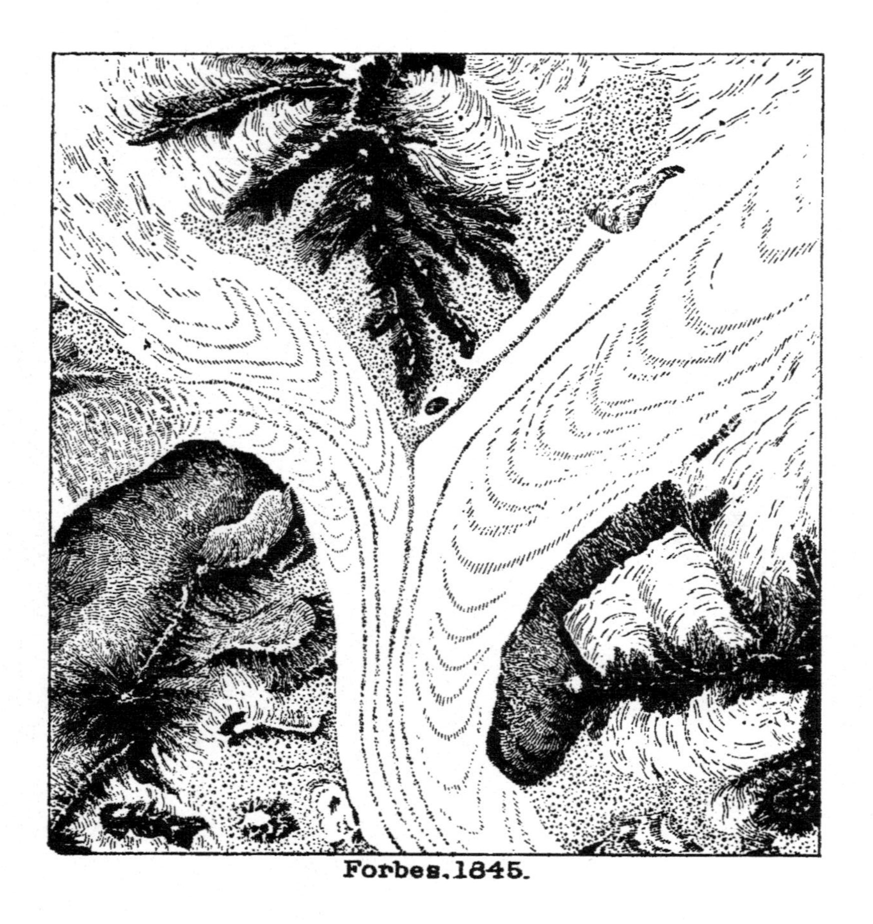

Forbes. 1845.

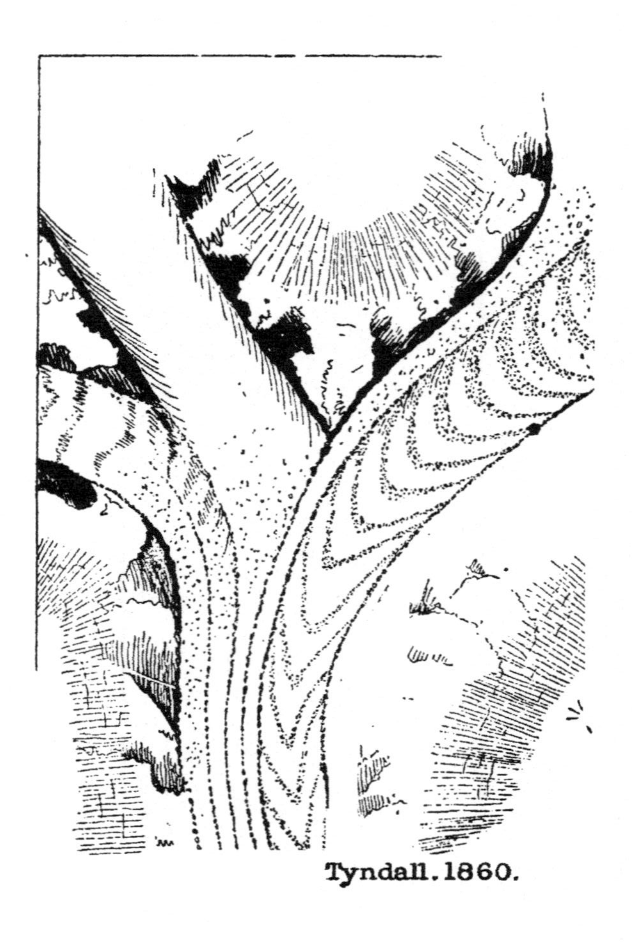

Tyndall. 1860.

Difficult enough, if approached even with a clear head, and open heart; acceptant of all help from former observers, and of all hints from nature and heaven; but very totally insoluble, when approached by men whose poor capacities for original thought are unsteadied by conceit, and paralyzed by envy.

11. In my second plate, I have given, side by side, a reduction, to half-scale, of part of Forbes's exquisite chart of the Mer de Glace, published in 1845, from his own survey made in 1842; and a reproduction, approximately in facsimile, of Professor Tyndall's woodcut, from his own 'eye-sketch' of the same portion of the glacier "as seen from the cleft station, Trélaporte," published in 1860.*

That Professor Tyndall is unable to draw anything as seen from anywhere, I observe to be a matter of much self-congratulation to him; such inability serving farther to establish the sense of his proud position as a man of science, above us poor artists, who labour under the disadvantage of being able with some accuracy to see, and with some fidelity to represent, what we wish to talk about. But when he found himself so resplendently inartistic, in the eye-sketch in question, that the expression of his scientific vision became, for less scientific persons, only a very bad

* 'Glaciers of the Alps,' p. 369. Observe also that my engraving, in consequence of the reduced scale, is grievously inferior to Forbes's work; but quite effectually and satisfactorily reproduces Professor Tyndall's, of the same size as the original.

map, it was at least incumbent on his Royally-social Emi nence to ascertain whether any better map of the same places had been published before. And it is indeed clear, in other places of his book, that he was conscious of the existence of Forbes's chart; but did not care to refer to it on this occasion, because it contained the correction of a mistake made by Forbes in 1842, which Professor Tyndall wanted, himself, to have the credit of correcting; leaving the public at the same time to suppose it had never been corrected by its author.

12. This manner, and temper, of reticence, with its relative personal loquacity, is not one in which noble science can be advanced; or in which even petty science can be increased. Had Professor Tyndall, instead of seeking renown by the exposition of Forbes's few and minute mistakes, availed himself modestly of Forbes's many and great discoveries, ten years of arrest by futile discussion and foolish speculation might have been avoided in the annals of geology; and assuredly it would not have been left for a despised artist to point out to you, this evening, the one circumstance of importance in glacier structure which Forbes has not explained.

13. You may perhaps have heard I have been founding my artistic instructions lately on the delineation of a jam-pot. Delighted by the appearance of that instructive object, in the Hotel du Mont Blanc, at St. Martin's, full of Chamouni honey, of last year, stiff and white, I found it also gave me command of the best pos-

sible material for examination of glacial action on a small scale.

Pouring a little of its candied contents out upon my plate, by various tilting of which I could obtain any rate of motion I wished to observe in the viscous stream; and encumbering the sides and centre of the said stream with magnificent moraines composed of crumbs of toast, I was able, looking alternately to table and window, to compare the visible motion of the mellifluous glacier, and its transported toast, with the less traceable, but equally constant, motion of the glacier of Bionnassay, and its transported granite. And I thus arrived at the perception of the condition of glacial structure, which though, as I told you just now, not, I believe, hitherto illustrated, it is entirely in your power to illustrate for yourselves in the following manner.

If you will open a fresh pot of honey to-morrow at breakfast, and take out a good table-spoonful of it, you will see, of course, the surface generally ebb in the pot. Put the table-spoonful back in a lump at one side, and you will see the surface generally flow in the pot. The lump you have put on at the side does not diffuse itself over the rest; but it sinks into the rest, and the entire surface rises round it, to its former level.

Precisely in like manner, every pound of snow you put on the top of Mont Blanc, eventually makes the surface of the glaciers rise at the bottom.*

* Practically hyperbolic expression, but mathematically true.

15. That is not impulsive action, mind you. That is mere and pure viscous action—the communication of force equally in every direction among slowly moving particles. I once thought that this force might also be partially elastic, so that whereas, however vast a mass of honey you had to deal with,—a Niagara of honey,—you never could get it to leap like a sea-wave at rocks, ice might yet, in its fluency, retain this power of leaping; only slowly,—taking a long time to rise. yet obeying the same mathematic law of impulse as a sea-breaker; but ascending through æras of surge, and communicating, through æras, its recoil. The little ripple of the stream breaks on the shore,—quick, quick, quick. The Atlantic wave slowly uplifts itself to its plunge, and slowly appeases its thunder. The ice wave—if there be one—would be to the Atlantic wave as the ocean is to the brook.

If there be one! The question is of immense—of vital—importance, to that of glacier action on crag: but, before attacking it, we need to know what the lines of motion are,—first, in a subsiding table-spoonful of honey; secondly, in an uprearing Atlantic wave; and, thirdly, in the pulsatory festoons of a descending cataract, obtained by the *relaxation* of its mass, while the same pulsatory action is displayed, as unaccountably, by a glacier cataract,* in the *compression* of its mass.

---

* Or a stick of sealing-wax. Warm one at the fire slowly through; and bend it into the form of a horseshoe. You will then see, through

And, on applying to learned men in Oxford and Cambridge * for elucidation of these modes of motion, I find that, while they can tell me everything I don't want to know, about the collision and destruction of planets, they are not entirely clear on the subject either of the diffusion of a drop of honey from its comb, or the confusion of a rivulet among its cresses. Of which difficult matters, I will therefore reserve inquiry to another chapter; anticipating, however, its conclusions, for the reader's better convenience, by the brief statement, that glacier ice has no power of springing whatever;—that it cannot descend into a rock-hollow, and sweep out the bottom of it, as a cascade or a wave can; but must always sluggishly fill it to the brim before flowing over; and accumulate, beneath, under dead ice, quiet as the depths of a mountain tarn, the fallen ruins of its colossal shore.

---

a lens of moderate power, the most exquisite facsimiles of glacier fissure produced by extension, on its convex surface, and as faithful image of glacier surge produced by compression, on its concave one.

In the course of such extension, the substance of the ice is actually expanded, (see above, Chap. IV., § 7,) by the widening of every minute fissure; and in the course of such compression, reduced to apparently solid ice, by their closing. The experiments both of Forbes and Agassiz appear to indicate that the original fissures are never wholly effaced by compression; but I do not myself know how far the supposed result of these experiments may be consistent with ascertained phenomena of regelation.

* I have received opportune and kind help, from the other side of the Atlantic waves, in a study of them by my friend Professor Rood.

## CHAPTER VII.

### THE IRIS OF THE EARTH.

*Lecture given at the London Institution, February* 17*th and March* 28*th*, 1876,*—*the subject announced being*, "AND THE GOLD OF THAT LAND IS GOOD: THERE IS BDELLIUM AND THE ONYX STONE.

1. THE subject which you permit me the pleasure of illustrating to you this evening, namely, the symbolic use of the colours of precious stones in heraldry, will, I trust, not interest you less because forming part both of the course of education in art which I have been permitted to found in Oxford; and of that in physical science, which I am about to introduce in the Musuem for working men at Sheffield.

I say 'to introduce,' not as having anything novel to teach, or show; for in the present day I think novelty the worst enemy of knowledge, and my introductions are only of things forgotten. And I am compelled to be pertina-

---

* The abrupt interpolation of this lecture in the text of Deucalion is explained in the next chapter.

ciously—it might even seem, insolently, separate in effort from many who would help me, just because I am resolved that no pupil of mine shall see anything, or learn, but what the consent of the past has admitted to be beautiful, and the experience of the past has ascertained to be true. During the many thousand years of this world's existence the persons living upon it have produced more lovely things than any of us can ever see; and have ascertained more profitable things than any of us can ever know. Of these infinitely existing, beautiful things, I show to my pupils as many as they can thoroughly see,—not more; and of the natural facts which are positively known, I urge them to know as many as they can thoroughly know, —not more; and absolutely forbid all debate whatsoever. The time for debate is when we have become masters—not while we are students. And the wisest of masters are those who debate least.

2. For my own part—holding myself nothing better than an advanced student, guiding younger ones,—I never waste a moment of life in dispute, or discussion. It is at least ten years since I ceased to speak of anything but what I had ascertained; and thus becoming, as far as I know, the most practical and positive of men, left discourse of things doubtful to those whose pleasure is in quarrel;—content, for my pupils and myself, to range all matters under the broad head of things certain, with which we are vitally concerned, and things uncertain, which don't in the least matter

3. In the working men's museum at Sheffield, then, I mean to place illustrations of entirely fine metal-work, including niello and engraving; and of the stones, and the Flora and Fauna, of Yorkshire, Derbyshire, Durham, and Westmoreland;* together with such foreign examples as may help to the better understanding of what we have at home. But in teaching metal-work, I am obliged to exhibit, not the uses of iron and steel only, but those also of the most precious metals, and their history; and for the understanding of any sort of stones, I must admit precious stones, and their history. The first elements of both these subjects, I hope it may not be uninteresting to you to follow out with me this evening.

4. I have here, in my right hand, a little round thing, and in my left a little flat one, about which, and the like of them, it is my first business to explain, in Sheffield, what may *positively* be known. They have long been both, to me, subjects of extreme interest; and I do not hesitate to say that I know more about them than most people: but that, having learned what I can, the happy feeling of wonder is always increasing upon me—how little that is! What an utter mystery both the little things still are!

5. This first—in my right hand—is what we call a 'pebble,'† or rolled flint, presumably out of Kensington

---

* Properly, Westmereland, the district of Western Meres.

† I. A. I. Sheffield Museum; see Chapter VIII.

gravel-pits. I picked it up in the Park,—the first that lay loose, inside the railings, at the little gate entering from Norfolk Street. I shall send it to Sheffield; knowing that like the bit of lead picked up by Saadi in the 'Arabian Nights,' it will make the fortune of Sheffield, scientifically, —if Sheffield makes the most of it, and thoroughly learns what it is.

6. What it *is*, I say,—you observe;—not merely, what it is *made of*. Anybody—the pitifullest apothecary round the corner, with a beggarly account of empty boxes—can tell you that. It is made of brown stuff called silicon, and oxygen, and a little iron; and so any apothecary can tell what you all who are sitting there are made of:—you, and I, and all of us, are made of carbon, nitrogen, lime, and phosphorus, and seventy per cent. or rather more of water; but then, that doesn't tell us what we are,—what a child is, or what a boy is,—much less what a man is,—least of all, what supremely inexplicable woman is. And so, in knowing only what it is made of, we don't know what a flint is.

7. To know what it IS, we must know what it can do, and suffer.

That it can strike steel into white-hot fire, but can itself be melted down like water, if mixed with ashes; that it is subject to laws of form one jot of which it cannot violate, and yet which it can continually evade, and apparently disobey; that in the fulfilment of these it becomes pure,—in rebellion against them, foul and base;

that it is appointed on our island coast to endure for countless ages, fortifying the sea cliff; and on the brow of that very cliff, every spring, to be dissolved, that the green blades of corn may drink it with the dew;—that in its noblest forms it is still imperfect, and in the meanest, still honourable,—this, if we have rightly learned, we begin to know what a flint is.

8. And of this other thing, in my left hand,—this flat bit of yellow mineral matter,—commonly called a 'sovereign,' not indeed to be picked up so easily as the other —(though often, by rogues, with small pains;)—yet familiar enough to the sight of most of us, and *too* familiar to our thought,—there perhaps are the like inquiries to be put. What *is* it? What can it do; and for whom? This shape given to it by men, bearing the image of a Cæsar;—how far does this make it a thing which is Cæsar's? the opposed image of a saint, riding against a dragon—how far does this make it a thing which is of Saints? Is its testimony true, or conceivably true, on either side? Are there yet Cæsars ruling us, or saints saving us, to whom it does of right belong?

9. And the substance of it,—not separable, this, into others, but a pure element,—what laws are over it, other than Cæsar's; what forms must it take, of its own, in eternal obedience to invisible power, if it escape our human hammer-stroke? How far, in its own shape, or in this, is it itself a Cæsar; inevitable in authority; secure of loyalty, loveable, and meritorious of love? For, read-

ing its past history, we find it has been much beloved, righteously or iniquitously,—a thing to be known the grounds of, surely?

10. Nay, also of this dark and despised thing in my right hand, we must ask that higher question, has it ever been beloved? And finding in its past history that in its pure and loyal forms, of amethyst, opal, crystal, jasper, and onyx, *it* also has been much beloved of men, shall we not ask farther whether it deserves to be beloved,—whether in wisdom or folly, equity or inequity, we give our affections to glittering shapes of clay, and found our fortunes on fortitudes of stone; and carry down from lip to lip, and teach, the father to the child, as a sacred tradition, that the Power which made us, and preserves, gave also with the leaves of the earth for our food, and the streams of the earth for our thirst, so also the dust of the earth for our delight and possession: bidding the first of the Rivers of Paradise roll stainless waves over radiant sands, and writing, by the word of the Spirit, of the Rocks that it divided, "The gold of that land is good; there also is the crystal, and the onyx stone."

11. Before I go on, I must justify to you the familiar word I have used for the rare one in the text.

If with mere curiosity, or ambitious scholarship, you were to read the commentators on the Pentateuch, you might spend, literally, many years of life, on the discussions as to the kinds of the gems named in it; and be no wiser at the end than you were at the beginning. But if, honestly and earnestly desiring to know the meaning of the book itself, you set yourself to read with such ordinary help as a good concordance and dictionary, and with fair knowledge of the two languages in which the Testaments have been clearly given to us, you may find out all you need know, in an hour.

12. The word 'bdellium' occurs only twice in the Old Testament: here, and in the book of Numbers, where you are told the manna was of the colour or look of bdellium. There, the Septuagint uses for it the word κρύσταλλος, crystal, or more properly anything congealed by cold; and in the other account of the manna, in Exodus, you are told that, after the dew round the camp was gone up, "there lay a small round thing—as small as the *hoar-frost* upon the ground." Until I heard from my friend Mr. Tyrrwhitt * of the cold felt at night in camping on Sinai, I could not understand how deep the feeling of the Arab, no less than the Greek, must have been respecting the divine gift of the dew,—nor with what sense of thankfulness for miraculous blessing the question of Job would be uttered, "The hoary frost of heaven, who hath gendered it?" Then compare the first words of the blessing of Isaac: "God give thee of the dew of heaven, and of

* See some admirable sketches of travelling in the Peninsula of Sinai, by this writer, in 'Vacation Tourists,' Macmillan, 1864. "I still remember," he adds in a private letter to me, "that the frozen towels stood on their edges as stiff as biscuits. By 11 A.M. the thermometer had risen to 85°, and was still rising."

the fatness of earth;" and, again, the first words of the song of Moses: "Give ear, oh ye heavens,—for my speech shall distil as the dew;" and you will see at once why this heavenly food was made to shine clear in the desert, like an enduring of its dew;—Divine remaining for continual need. Frozen, as the Alpine snow—pure for ever.

13. Seize firmly that first idea of the manna, as the type of the bread which is the Word of God; * and then look on for the English word 'crystal' in Job, of Wisdom, "It cannot be valued with the gold of Ophir, with the precious *onyx*, or the sapphire: the *gold and the crystal* shall not equal it, neither shall it be valued with pure gold;" in Ezekiel, "firmament of the terrible crystal," or in the Apocalypse, "A sea of glass, like unto crystal,—water of life, clear as crystal"—"light of the city like a stone most precious, even like a jasper stone, clear as crystal." Your understanding the true meaning of all these passages depends on your distinct conception of the permanent clearness and hardness of the Rock-crystal. You may trust me to tell you quickly, in this matter, what you may all for yourselves discover if you will read.

14. The three substances named here in the first account of Paradise, stand generally as types—the Gold

---

* Sir Philip Sidney, in his translation of the ἄρτον οὐρανοῦ of the 105th Psalm, completes the entire range of idea,

"Himself, from skies, their hunger to repel,
*Candies* the grasse with sweete congealed dew."

of all precious metals; the CRYSTAL of all clear precious stones prized for *lustre;* the ONYX of all opaque precious stones prized for *colour.* And to mark this distinction as a vital one,—in each case when the stones to be set for the tabernacle-service are named, the onyx is named separately. The Jewish rulers brought "onyx stones, and stones to be set for the ephod, and for the breastplate."* And the onyx is used thrice, while every other stone is used only once, in the High Priest's robe; two onyxes on the shoulders, bearing the twelve names of the tribes, six on each stone, (Exod. xxviii. 9, 10,) and one in the breast-plate, with its separate name of one tribe, (Exod. xxviii. 20.)

15. A. Now note the importance of this grouping. The Gold, or precious metal, is significant of all that the power of the beautiful earth, gold, and of the strong earth, iron, has done for and against man. How much evil I need not say. How much good is a question I will endeavour to show some evidence on forthwith.

B. The Crystal is significant of all the power that jewels, from diamonds down through every Indian gem to the glass beads which we now make for ball-dresses, have had over the imagination and economy of men and women—from the day that Adam drank of the water of the crystal river to this hour.

* Exod. xxv. 7, xxxv. 27, comparing Job above quoted, and Ezekiel xxviii. 13.

How much evil that is, you partially know; how much good, we have to consider.

c. The Onyx is the type of all stones arranged in bands of different colours; it means primarily, nail-stone—showing a separation like the white half-crescent at the root of the finger-nail; not without some idea of its subjection to laws of life. Of these stones, part, which are flinty, are the material used for cameos and all manner of engraved work and pietra dura; but in the great idea of banded or belted stones, they include the whole range of marble, and especially alabaster, giving the name to the alabastra, or vases used especially for the containing of precious unguents, themselves more precious;* so that this stone, as best representative of all others, is chosen to be the last gift of men to Christ, as gold is their first; incense with both: at His birth, gold and frankincense; at His death, alabaster and spikenard.

16. The two sources of the material wealth of all nations were thus offered to the King of men in their simplicity. But their power among civilized nations has been owing to their workmanship. And if we are to ask whether the gold and the stones are to be holy, much more have we to ask if the worker in gold, and the worker in stone, are to be conceived as exercising holy function.

17. Now, as we ask of a stone, to know what it is, what

* Compare the "Nardi parvos onyx," which was to be Virgil's feast-gift, in spring, to Horace.

it can do, or suffer, so of a human creature, to know what it is, we ask what it can do, or suffer.

So that we have two scientific questions put to us, in this matter: how the stones came to be what *they* are—or the law of Crystallization; and how the jewellers came to be what *they* are—or the law of Inspiration. You see how vital this question is to me, beginning now actually to give my laws of Florentine art in English Schools! How can artists be made artists,—in gold and in precious stones? whether in the desert, or the city?—and if in the city, whether, as at Jerusalem, so also in Florence, Paris, or London?

Must we at this present time, think you, order the jewellers, whom we wish to teach, merely to study and copy the best results of past fashion? or are we to hope that some day or other, if we behave rightly, and take care of our jewels properly, we shall be shown also how to set them; and that, merely substituting modern names for ancient ones, some divine message will come to our craftsmen, such as this: 'See, I have called by name Messrs. Hunt and Roskell, and Messrs. London and Ryder, and I have filled them with the Spirit of God, in wisdom and in understanding, and in all manner of workmanship, to work in gold, and in silver, and in brass, and in cutting of stones,'?

18. This sentence, which, I suppose, becomes startling to your ear in the substitution of modern for ancient names, is the first, so far as I know, distinctly referring

to the ancient methods of instruction in the art of jewel lery. So also the words which I have chosen for the title (or, as perhaps some of my audience may regretfully think it should be called, the text,) of my lecture, are the first I know that give any account of the formation or existence of jewels. So that the same tradition, whatever its value, which gave us the commands we profess to obey for our moral law, implies also the necessity of inspired instruction for the proper practice of the art of jewellery; and connects the richness of the earth in gold and jewels with the pleasure of Heaven that we should use them under its direction. The scientific mind will of course draw back in scorn from the idea of such possibility; but then, the scientific mind can neither design, itself, nor perceive the power of design in others. And practically you will find that all noble design in jewellery whatsoever, from the beginning of the world till now, has been either instinctive,—done, that is to say, by tutorship of nature, with the innocent felicity and security of purely animal art,—Etruscan, Irish, Indian, or Peruvian gold being interwoven with a fine and unerring grace of industry, like the touch of the bee on its cell and of the bird on her nest,—or else, has been wrought into its finer forms, under the impulse of religion in sacred service, in crosier, chalice, and lamp; and that the best beauty of its profane service has been debased from these. And the three greatest masters of design in jewellery, the 'facile principes' of the entire European School, are—centrally, the one who definitely worked always with appeal for inspiration—Angelico of Fésole; and on each side of him, the two most earnest reformers of the morals of the Christian Church—Holbein, and Sandro Botticelli.

19. I have first answered this, the most close home of the questions,—how men come to be jewellers. Next, how do stones come to be jewels? It seems that by all religious, no less than all profane, teaching or tradition, these substances are asserted to be precious,—useful to man, and sacred to God. Whether we have not made them deadly instead of useful,—and sacrificed them to devils instead of God,—you may consider at another time. To-night, I would examine only a little way the methods in which they are prepared by nature, for such service as they are capable of.

20. There are three great laws by which they, and the metals they are to be set in, are prepared for us; and at present all these are mysteries to us.

I. The first, the mystery by which "surely there is a vein for the silver, and a place for the gold whence * they fine it." No geologist, no scientific person whatsoever, can tell you how this gold under my hand was brought into this cleft in the bdellium; † no one knows where it was before, or how it got here: one thing only seems to be manifest—that it was not here always. This white

* 'Whence,' not 'where,' they sift or wash it: ὅθεν διηθεῖται, LXX.

† 20. A. 1. Sheffield Museum.

bdellium itself closes rents, and fills hollows, in rocks which had to be rent before they could be rejoined, and hollowed before they could be refilled. But no one hitherto has been able to say where the gold first was, or by what process it came into this its resting-place. First mystery, then,—that there is a vein for the silver, and a place for the gold.

II. The second mystery is that of crystallization; by which, obeying laws no less arbitrary than those by which the bee builds her cell—the water produced by the sweet miracles of cloud and spring freezes into the hexagonal stars of the hoar-frost;—the flint, which can be melted and diffused like water, freezes also, like water, into *these* hexagonal towers of everlasting ice; * and the clay, which can be dashed on the potter's wheel as it pleaseth the potter to make it, can be frozen by the touch of Heaven into the hexagonal star of Heaven's own colour—the sapphire.

III. The third mystery, the gathering of crystals themselves into ranks or bands, by which Scotch pebbles are made, not only is at present unpierced, but—which is a wonderful thing in the present century—it is even untalked about. There has been much discussion as to the nature of metallic veins; and books have been written with indefatigable industry, and splendid accumulation of facts, on the limits, though never on the methods, of crystallization. But of the structure of banded stones not

* I. Q. 11. Sheffield Museum.

a word is ever said, and, popularly, less than nothing known; there being many very false notions current respecting them, in the minds even of good mineralogists.

And the basis of what I find to be ascertainable about them, may be told with small stress to your patience.

21. I have here in my hand,* a pebble which used to decorate the chimney-piece of the children's playroom in my aunt's house at Perth, when I was seven years old, just half a century ago; which pebble having come out of the hill of Kinnoull, on the other side of the Tay, I show you because I know so well where it came from, and can therefore answer for its originality and genuineness.

22. The hill of Kinnoull, like all the characteristic crags or craigs of central Scotland, is of a basaltic lava—in which, however, more specially than in most others, these balls of pebble form themselves. And of these, in their first and simplest state, you may think as little pieces of flint jelly, filling the pores or cavities of the rock.

Without insisting too strictly on the analogy—for Nature is so various in her operations that you are sure to be deceived if you ever think one process has been in all respects like another—you may yet in most respects think of the whole substance of the rock as a kind of brown bread, volcanically baked, the pores and cavities of which, when it has risen, are filled with agate or onyx jelly, as the sim-

* I. A. 8. Sheffield Museum.

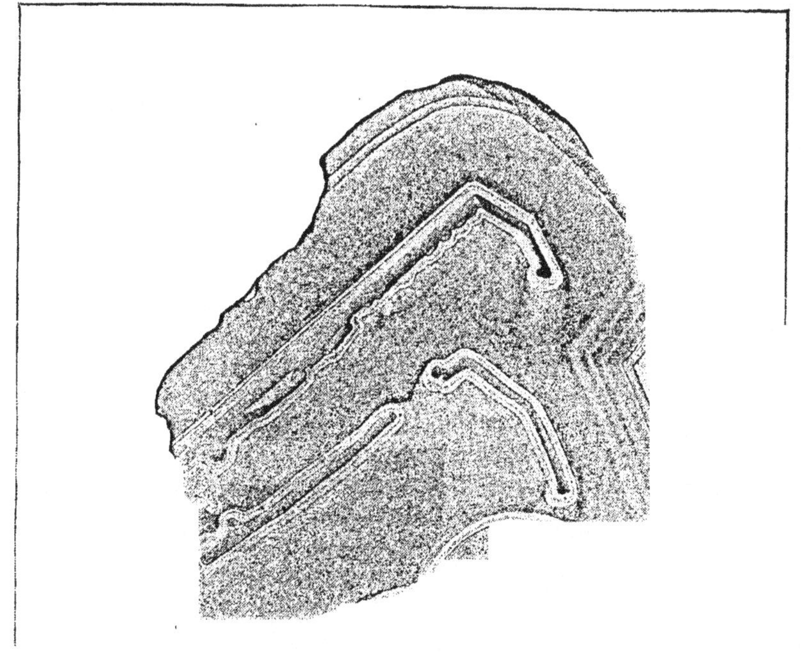

ilar pores of a slice of quartern loaf are filled with butter, if the cook has spread it in a hurry.

23. I use this simile with more satisfaction, because, in the course of last autumn, I was making some practical experiments on glacial motion—the substances for experiment being supplied to me in any degree of congelation or regelation which might be required, by the perfectly angelic cook of a country friend, who not only gave me the run of her kitchen, but allowed me to make domical mountains of her best dish-covers, and tortuous valleys of her finest napkins;—under which altogether favourable conditions, and being besides supplied with any quantity of ice-cream and blancmange, in every state of frost and thaw, I got more beautiful results, both respecting glacier motion, and interstratified rocks, than a year's work would have reached by unculinary analysis. Keeping, however —as I must to-night—to our present question, I have here a piece of this baked volcanic rock, which is as full of agate pebbles as a plum-pudding is of currants; each of these agate pebbles consisting of a clear green chalcedony, with balls of banded agate formed in the midst, or at the sides of them. This diagram * represents one enlarged.

And you have there one white ball of agate, floating apparently in the green pool, and a larger ball, which is cut through by the section of the stone, and shows you the banded structure in the most exquisite precision.

---

* This drawing is in Sheffield Museum.

24. Now, there is no doubt as to the possible formation of these balls in melted vitreous substance as it cools, because we get them in glass itself, when gradually cooled in old glass-houses; and there is no more difficulty in accounting for the formation of round agate balls of this character than for that of common globular chalcedony. But the difficulty begins when the jelly is not allowed to remain quiet, but can run about while it is crystallizing. Then you get glutinous forms that choke cavities in the rock, in which the chalcedony slowly runs down the sides, and forms a level lake at the bottom; and sometimes you get the whole cavity filled with lake poured over lake, the liquid one over the frozen, floor and walls at last encrusted with onyx fit for kings' signets.*

25. Of the methods of engraving this stone, and of its general uses and values in ancient and modern days, you will find all that can interest you, admirably told by Mr. King, in his book on precious stones and gems, to which I owe most of the little I know myself on this subject.

To-night, I would only once more direct your attention to that special use of it in the dress of the Jewish High Priest; that while, as one of the twelve stones of the breastplate, it was engraved like the rest with the name of a single tribe, two larger onyxes were used for the shoulder-studs of the ephod; and on these, the names of

---

* I am obliged to omit here the part of the lecture referring to diagrams. It will be given in greater detail in the subsequent text.

all the twelve tribes were engraved, six upon each. I do not infer from this use of the onyx, however, any pre-eminence of value, or isolation of symbolism, in the stone; I suppose it to have been set apart for the more laborious piece of engraving, simply because larger surfaces of it were attainable than of true gems, and its substance was more easily cut. I suppose the bearing of the names on the shoulder to be symbolical of the priest's sacrificial office in bearing the guilt and pain of the people; while the bearing of them on the breast was symbolical of his pastoral office in teaching them: but, except in the broad distinction between gem and onyx, it is impossible now to state with any certainty the nature or meaning of the stones, confused as they have been by the most fantastic speculation of vain Jewish writers themselves.

There is no such difficulty when we pass to the inquiry as to the use of these stones in Christian Heraldry, on the breastplate and shield of the Knight; for that use is founded on natural relations of colour, which cannot be changed, and which will become of more and more importance to mankind in proportion to the degree in which Christian Knighthood, once proudly faithful to Death, in War, becomes humbly faithful to Life, in Peace.

27. To these natural relations of colour, the human sight, in health, is joyfully sensitive, as the ear is to the harmonies of sound; but what healthy sight is, you may well suppose, I have not time to define to-night;—the nervous power of the eye, and its delight in the pure hues

of colour presented either by the opal, or by wild flowers, being dependent on the perfect purity of the blood supplied to the brain, as well as on the entire soundness of the nervous tissue to which that blood is supplied. And how much is required, through the thoughts and conduct of generations, to make the new blood of our race of children pure—it is for your physicians to tell you, when they have themselves discovered this medicinal truth, that the divine laws of the life of Men cannot be learned in the pain and death of Brutes.

28. The natural and unchangeable system of visible colour has been lately confused, in the minds of all students, partly by the pedantry of unnecessary science; partly by the formalism of illiberal art: for all practical service, it may be stated in a very few words, and expressed in a very simple diagram.

28. There are three primary colours, Red, Blue, and Yellow; three secondary, formed by the union of any two of these; and one tertiary, formed by the union of all three.

If we admitted, as separate colours, the different tints produced by varying proportions of the composing tints, there would of course be an infinite number of secondaries, and a wider infinitude of tertiaries. But tints can be systematically arranged only by the elements of them, not the proportions of those elements. Green is only green, whether there be less or more of blue in it; purple only purple, whether there be less or more of red in it; scarlet only scarlet, whether there be less or more of yellow in it; and the tertiary gray only gray, in whatever proportions the three primaries are combined in it.

29. The diagram used in my drawing schools to express the system of these colours will be found coloured in the 'Laws of Fésole':—this figure will serve our present purpose.*

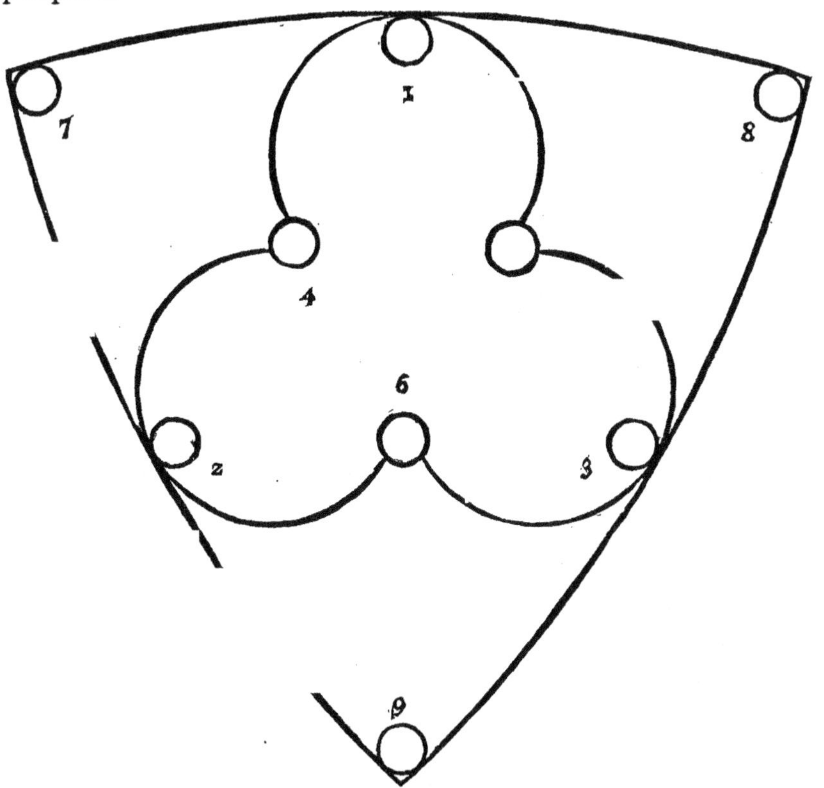


The simple trefoil produced by segments of three cir-

* Readers interested in this subject are sure to be able to enlarge and colour it for themselves. I take no notice of the new scientific theories of primary colour: because they are entirely false as applied to practical work, natural or artistic. Golden light in blue sky makes green sky; but green sky and red clouds can't make yellow sky.

cles in contact, is inscribed in a curvilinear equilateral triangle. Nine small circles are set,—three in the extremities of the foils, three on their cusps, three in the angles of the triangle.

The circles numbered 1 to 3 are coloured with the primitive colours; 4 to 6, with the secondaries; 7 with white; 8 with black; and the 9th, with the tertiary, gray.

30. All the primary and secondary colours are capable of infinitely various degrees of intensity or depression: they pass through every degree of increasing light, to perfect light, or white; and of increasing shade, to perfect absence of light, or black. And these are essential in the harmony required by sight; so that no group of colours can be perfect that has not white in it, nor any that has not black; or else the abatement or modesty of them, in the tertiary, gray. So that these three form the limiting angles of the field, or cloudy ground of the rainbow. "I do set my bow in the cloud."

And the nine colours of which you here see the essential group, have, as you know, been the messenger Iris; exponents of the highest purpose, and records of the perfect household purity and honour of men, from the days when Hesiod blazoned the shield of Heracles, to the day when the fighting Temeraire led the line at Trafalgar,—the Victory following her, with three flags nailed to her masts, for fear one should be shot away.

31. The names of these colours in ordinary shields

of knighthood, are those given opposite, in the left-hand column. The names given them in blazoning the shields of nobles, are those of the correspondent gems: of heraldry by the planets, reserved for the shields of kings, I have no time to speak, to-night, except incidentally.

A. THE PRIMARY COLOURS.

| | | |
|---|---|---|
| 1. | Or. | Topaz. |
| 2. | Gules. | Ruby |
| 3. | Azure. | Sapphire. |

B. THE SECONDARY COLOURS.

| | | |
|---|---|---|
| 4. | Écarlate. | Jasper. |
| 5. | Vert. | Emerald. |
| 6. | Purpure. | Hyacinth. |

C. THE TERTIARY COLOURS.

| | | |
|---|---|---|
| 7. | Argent. | Carbuncle. |
| 8. | Sable. | Diamond. |
| 9. | Colombin. | Pearl. |

32. I. Or. Stands between the light and darkness; as the sun, who "rejoiceth as a strong man to run his course," between the morning and the evening. Its heraldic name, in the shields of kings, is Sol: the Sun, or Sun of Justice; and it stands for the strength and honour of all men who run their race in noble work; whose path "is as the shining light, that shineth more and more unto the perfect day."

For theirs are the works which are to shine before men, that they may glorify our Father. And they are also to shine before God, so that with respect to them, what was written of St. Bernard may be always true: "Opera sancti patris velut Sol in conspectu Dei."

For indeed they are a true light of the world, infinitely more good, in the sight of its Creator, than the dead flame of its sunshine; and the discovery of modern science, that all mortal strength is from the sun, which has thrown irrational persons into stupid atheism, as if there was no God but the sun, is indeed the accurate physical expression of this truth, that men, rightly active, are living sunshine.

II. Gules, (rose colour,) from the Persian word 'gul,' for the rose. It is the exactly central hue between the dark red, and pale red, or wild-rose. It is the colour of love, the fulfilment of the joy and of the love of life upon the earth. And it is doubly marked for this symbol. We saw earlier, how the vase given by the Madelaine was precious in its material; but it was also to be indicated as precious in its form. It is not only the substance, but the form of the Greek urn, which gives it nobleness; and these vases for precious perfume were tall, and shaped like the bud of the rose. So that the rose-bud itself, being a vase filled with perfume, is called also 'alabastron'; and Pliny uses that word for it in describing the growth of the rose.

The stone of it is the Ruby.

III. Azure. The colour of the blue sky in the height

of it, at noon;—type of the fulfilment of all joy and love in heaven, as the rose-colour, of the fulfilment of all joy and love in earth. And the stone of this is the Sapphire; and because the loves of Earth and Heaven are in truth one, the ruby and sapphire are indeed the same stone; and they are coloured as if by enchantment,—how, or with what, no chemist has yet shown,—the one azure, and the other rose.

And now you will understand why, in the vision of the Lord of Life to the Elders of Israel, of which it is written, "Also they saw God, and did eat and drink," you are told, "Under His feet was a plinth of sapphire, and, as it were, the body of Heaven in its clearness."

IV. Écarlate (scarlet). I use the French word, because all other heraldic words for colours are Norman-French. The ordinary heraldic term here is 'tenné' (tawny); for the later heralds confused scarlet with gules; but the colour first meant was the sacred hue of human flesh—Carnation;—*in*carnation: the colour of the body of man in its beauty; of the maid's scarlet blush in noble love; of the youth's scarlet glow in noble war; the dye of the earth into which heaven has breathed its spirit: incarnate strength—incarnate modesty.

The stone of it is the Jasper, which as we shall see, is coloured with the same iron that colours the human blood; and thus you can understand why on the throne, in the vision of the returning Christ, "He that sat was to look upon like a jasper and a sardine stone."

V. Vert, (viridis,) from the same root as the words 'virtue' and 'virgin,'—the colour of the green rod in budding spring; the noble life of youth, born in the *spirit*,—as the scarlet means, the life of noble youth, in *flesh*.* It is seen most perfectly in clear air after the sun has set,—the blue of the upper sky brightening down into it. It is the true colour of the eyes of Athena,—Athena *Γλαυκῶπις*,† looking from the west.

The stone of it is the Emerald; and I must stay for a moment to tell you the derivation of that word.

Anciently, it did not mean our emerald, but a massive green marble, veined apparently by being rent asunder, and called, therefore, the Rent or Torn Rock.

Now, in the central war of Athena with the Giants, the sign of her victory was that the earth was rent, the power of it torn, and the graves of it opened. We know this is written for the sign of a greater victory than hers. And the word which Hesiod uses—the oldest describer of this battle—is twice over the same: the sea roared, the heavens thundered, the earth cried out in being rent,

---

* Therefore, the Spirit of Beatrice is dressed in green, over *scarlet*, (not rose;—observe this specially).

> "Sovra candido vel, cinta d' oliva
> Donna m' apparve sotto verde manto,
> Vestita di color di *fiamma* viva."

† Accurately described by Pausanias, 1, xiv., as of the colour of a green lake, from the Tritonian pool; compare again the eyes of Beatrice.

ἐσμαράγησε. From that word you have "the rent rock," —in Latin, smaragdus; in Latin dialect, smaraudus—softened into emeraudu, emeraude, emerald. And now you see why "there was a rainbow round about the throne in sight like unto an emerald."

VI. Purpure. The true purple of the Tabernacle, "blue, purple, and scarlet"—the kingly colour, retained afterwards in all manuscripts of the Greek Gospels; therefore known to us absolutely by its constant use in illumination. It is rose colour darkened or saddened with blue; the colour of love in noble or divine sorrow; borne by the kings, whose witness is in heaven, and their labour on the earth. Its stone is the Jacinth, Hyacinth, or Amethyst,—"like to that sable flower inscribed with woe."

In these six colours, then, you have the rainbow, or angelie iris, of the light and covenant of life.

But the law of the covenant is, "I do set my bow in the cloud, on the shadow of death—and the ordinance of it."

And as here, central, is the sun in his strength, so in the heraldry of our faith, the morning and the evening are the first day,—and the last.

VII. Argent. Silver, or snow-colour; of the hoar-frost on the earth, or the star of the morning.

I was long hindered from understanding the entire group of heraldic colours, because of the mistake in our use of the word 'carbuncle.' It is not the garnet, but the same stone as the ruby and sapphire—only crystallized white, instead of red or blue. It is the white sapphire, showing the hexagonal star of its crystallization perfectly; and therefore it becomes an heraldic bearing as a star.

And it is the personal bearing of that Geoffrey Plantagenet, who married Maud the Empress, and became the sire of the lords of England, in her glorious time.

VIII. Sable, (sable, sabulum,) the colour of sand of the great hour-glass of the world, outshaken. Its stone is the diamond—never yet, so far as I know, found but in the sand.* It is the symbol at once of dissolution, and of endurance: darkness changing into light—the adamant of the grave.

IX. Gray. (When deep, the second violet, giving Dante's full chord of the seven colours.) The abatement of the light, the abatement of the darkness. Patience, between this which recedes and that which advances; the colour of the turtle-dove, with the message that the waters are abated; the colour of the sacrifice of the poor,—therefore of humility. Its stone is the Pearl; in Norman heraldry the Marguerite—the lowest on the shield, yet of great price; and because, through this virtue, open first the gates of Paradise, you are told that while the building of the walls of it was of jasper, every several gate was of one pearl.

33. You hear me tell you thus positively,—and without

---

* Or in rock virtually composed of it.

qualification or hesitation,—what these things mean. But mind, I tell you so, after thirty years' work, and that directed wholly to the one end of finding out the truth, whether it was pretty or ugly to look in face of. During which labour I have found that the ultimate truth, the central truth, is always pretty; but there is a superficial truth, or half-way truth, which may be very ugly; and which the earnest and faithful worker has to face and fight, and pass over the body of,—feeling it to be his enemy; but which a careless seeker may be stopped by, and a misbelieving seeker will be delighted by, and stay with, gladly.

34. When I first gave this lecture, you will find the only reports of it in the papers, with which any pains had been taken, were endeavours to make you disbelieve it, or misbelieve it,—that is to say, to make 'meseroyants' or 'miscreants' of you.

And among the most earnest of these, was a really industrious essay in the 'Daily Telegraph,'—showing evidence that the writer had perseveringly gone to the Heralds' Office and British Museum to read for the occasion; and, I think, deserving of serious notice because we really owe to the proprietors of that journal (who supplied the most earnest of our recent investigators with funds for his Assyrian excavations) the most important heraldic discoveries of the generations of Noah and Nimrod, that have been made since printing took the place of cuneiform inscription.

I pay, therefore, so much respect to the archæologians of Fleet Street as to notice the results of their suddenly stimulated investigations in heraldry.

35. "The lecturer appeared to have forgotten," they said, "that every nation had its own code of symbols, and that gules, or red, is denominated by the French heralds gueules, and is derived by the best French philologers from the Latin 'gula,' the gullet of a beast of prey."

It is perfectly true that the best French philologists do give this derivation; but it is also unfortunately true that the best French philologists are not heralds; and what is more, and worse, all modern heraldry whatsoever is, to the old science, just what the poor gipsy Hayraddin, in 'Quentin Durward,' is to Toison d'Or. But, so far from having 'forgotten,' as the writer for the press supposes I had, that there were knights of France, and Venice, and Florence, as well as England, it so happens that my first studies in heraldry were in *this* manuscript, which is the lesson-book of heraldry written for the young Archduke Charles of Austria; and in *this* one, which is a psalter written in the monastery of the Saint Chapelle for St. Louis, King of France; and on the upper page of which, here framed,* you will see written, in letters of gold, the record of the death of his mother, Blanche of Castile, on the 27th of November, next after St. Geneviève's day;

* The books referred to, in my rooms at Oxford, are always accessible for examination.

and on the under page, between the last lines of the Athanasian Creed, her bearing, the Castilian tower, alternating with the king's,—Azure, semé de France.

36. With this and other such surer authority than was open to the investigation of the press-writer, I will clear up for you his point about the word 'gules.' But I must go a long way back first. I do not know if, in reading the account of the pitching of the standards of the princes of Israel round the Tabernacle, you have ever been brought to pause by the singular covering given to the Tabernacle itself,—rams' skins dyed red, and *badgers'* skins. Of rams' skins, of course, any quantity could be had from the flocks, but of badgers', the supply must have been difficult!

And you will find, on looking into the matter, that the so-called badgers' skins were indeed those which young ladies are very glad to dress in at the present day,—sealskins; and that the meaning of their use in the Tabernacle was, that it might be adorned with the useful service of the *flocks* of the earth and sea: the multitude of the seals then in the Mediterranean being indicated to you both by the name and coinage of the city Phocæa; and by the attribution of them, to the God Proteus, in the first book of the Odyssey, under the precise term of flocks, to be counted by him as their shepherd.

37. From the days of Moses and of Homer to our own, the traffic in these precious wools and furs, in the Cashmere wool, and the fur, after the seal disappeared, of the

grey ermine, (becoming white in the Siberian winter,) has continued: and in the days of chivalry became of immense importance; because the mantle, and the collar fastening close about the neck, were at once the most useful and the most splendid piece of dress of the warrior nations, who rode and slept in roughest weather, and in open field. Now, these rams' skins, or fleeces, dyed of precious red, were continually called by their Eastern merchants 'the red things,' from the Zoroastrian word 'gul,'—taking the place of the scarlet Chlamydes, which were among the richest wealth of old Rome. The Latin knights could only render the eastern word 'gul' by gula; and so in St. Bernard's red-hot denunciation of these proud red dresses, he numbers chiefly among them the little red-dyed skins,—pelliculas rubricatas,—which they call gulæ: "Quas gulas vocant." These red furs, for wrist and neck, were afterwards supposed by bad Latinists to be called 'gulæ,' as *throat*-pieces. St. Bernard specifies them, also, in that office: "Even some of the clergy," he says, "have the red skins of weasels hanging from their necks—dependentes a collo"; this vulgar interpretation of gula became more commonly accepted, as intercourse with the East, and chivalric heraldry, diminished; and the modern philologist finally jumps fairly down the lion's throat, and supposes that the Tyrian purple, which had been the pride of all the Emperors of East and West, was named from a wild beast's gullet!

38. I do not hold for a mischance, or even for a chance

at all, that this particular error should have been unearthed by the hasty studies of the Daily Telegraph. It is a mistake entirely characteristic of the results of vulgar modern analysis; and I have exposed it in detail, that I might very solemnly warn you of the impossibility of arriving at any just conclusions respecting ancient classical languages, of which this heraldry is among the noblest, unless we take pains first to render ourselves *capable of the ideas* which such languages convey. It is perfectly true that every great symbol, as it has, on one side, a meaning of comfort, has on the other one of terror; and if to noble persons it speaks of noble things, to ignoble persons it will as necessarily speak of ignoble ones. Not under one only, but under all, of these heraldic symbols, as there is, for thoughtful and noble persons, the spiritual sense, so for thoughtless and sensual persons, there is the sensual one; and *can* be no other. Every word has only the meaning which its hearer can receive; you cannot express honour to the shameless, nor love to the unloving. Nay, gradually you may fall to the level of having words no more, either for honour or for love:

"There are whole nations," says Mr. Farrar, in his excellent little book on the families of speech, "people whom no nation now acknowledges as its kinsmen, whose languages, rich in words for all that can be eaten or handled, seem absolutely incapable of expressing the reflex conceptions of the intellect, or the higher forms of the consciousness; whose life seems confined to a gratification of animal wants, with no hope in the future, and no pride in the past. They are for the most part peoples without a literature, and without a history;—peoples whose tongues in some instances have twenty names for murder, but no name for love, no name for gratitude, no name for God."

39. The English nation, under the teaching of modern economists, is rapidly becoming one of this kind, which, deliberately living, not in love of God or man, but in defiance of God, and hatred of man, will no longer have in its heraldry, gules as the colour of love; but gules only as the colour of the throat of a wild beast. That will be the only part of the British lion symbolized by the British flag;—not the lion heart any more, but only the lion gullet.

And if you choose to interpret your heraldry in that modern fashion, there are volumes of instruction open for you everywhere. Yellow shall be to you the colour of treachery, instead of sunshine; green, the colour of putrefaction, instead of strength; blue, the colour of sulphurous hell-fire, instead of sunlit heaven; and scarlet, the colour of the harlot of Babylon, instead of the Virgin of God. All these are legitimate readings,—nay, inevitable readings. I said wrongly just now that you might choose what the symbols shall be to you. Even if you would, you cannot choose. They can only reflect to you what you have made your own mind, and can only herald to you what you have determined for your own fate.

40. And now, with safe understanding of the meaning of purple, I can show you the purple and dove-colour of St. Mark's, once itself a sea-borne vase of alabaster full of incense of prayers; and a purple manuscript,—floor, walls, and roof blazoned with the scrolls of the gospel.

They have been made a den of thieves, and these stones of Venice here in my hand * are rags of the sacred robes of her Church, sold, and mocked like her Master. They have parted her garments, and cast lots upon her vesture.

41. I return to our question at the beginning: Are we right in setting our hearts on these stones,—loving them, holding them precious?

Yes, assuredly; provided it is the stone we love, and the stone we think precious; and not ourselves we love, and ourselves we think precious. To worship a black stone, because it fell from heaven, may not be wholly wise, but it is half-way to being wise; half-way to worship of heaven itself. Or, to worship a white stone because it is dug with difficulty out of the earth, and to put it into a log of wood, and say the wood sees with it, may not be

---

* Portions of the alabaster of St. Mark's torn away for recent restorations. The destruction of the floor of the church, to give work to modern mosaic-mongers, has been going on for years. I cannot bear the pain of describing the facts of it, and must leave the part of the lecture referring to the colour of the marbles to be given farther on, in connection with some extracts from my 'Stones of Venice.' The superb drawing, by Mr. Bunney, of the north portico, which illustrated them, together with the alabasters themselves, will be placed in the Sheffield Museum.

wholly wise; but it is half-way to being wise; half-way to believing that the God who makes earth so bright, may also brighten the eyes of the blind. It is no true folly to think that stones see, but it *is*, to think that eyes do not; it is no true folly to think that stones live, but it *is*, to think that souls die; it is no true folly to believe that, in the day of the making up of jewels, the palace walls shall be compact of life above their corner-stone,—but it *is*, to believe that in the day of dissolution the souls of the globe shall be shattered with its emerald; and no spirit survive, unterrified, above the ruin.

42. Yes, pretty ladies! love the stones, and take care of them; but love your own souls better, and take care of *them*, for the day when the Master shall make up His jewels. See that it be first the precious stones of the breastplate of justice you delight in, and are brave in; not first the stones, of your own diamond necklaces * you delight in, and are fearful for, lest perchance the lady's maid miss that box at the station. Get your breastplate of truth first, and every earthly stone will shine in it.

* Do you think there was no meaning of fate in that omen of the diamond necklace; at the end of the days of queenly pride;—omen of another line, of scarlet, on many a fair neck? It was a foul story, you say—slander of the innocent. Yes, undoubtedly, fate meant it to be so. Slander, and lying, and every form of loathsome shame, cast on the innocently fading Royalty. For the corruption of the best is the worst; and these gems, which are given by God to be on the breast of the pure priest, and in the crown of the righteous king, sank into the black gravel of diluvium, under streams of innocent blood.

Alas! most of you know no more what justice means, than what jewels mean; but here is the pure practice of it to be begun, if you will, to-morrow.

43. For literal truth of your jewels themselves, absolutely search out and cast away all manner of false, or dyed, or altered stones. And at present, to make quite sure, wear your jewels uncut; they will be twenty times more interesting to you, so. The ruby in the British crown is uncut; and is, as far as my knowledge extends,—I have not had it to look at close,—the loveliest precious stone in the world. And, as a piece of true gentlewoman's and true lady's knowledge, learn to know these stones when you see them, uncut. So much of mineralogy the abundance of modern science may, I think, spare, as a piece of required education for the upper classes.

44. Then, when you know them, and their shapes, get your highest artists to design the setting of them. Holbein, Botticelli, or Angelico, will always be ready to design a brooch for you. Then you will begin to think how to get your Holbein and Botticelli, which will lead to many other wholesome thoughts.

45. And lastly, as you are true in the choosing, be just in the sharing, of your jewels. They are but dross and dust, after all; and you, my sweet religious friends, who are so anxious to impart to the poor your pearls of great price, may surely also share with them your pearls of little price. Strangely (to my own mind at least), you are not so zealous in distributing your estimable rubies, as you

are in communicating your *in*estimable wisdom. Of the grace of God, which you can give away in the quantity you think others are in need of, without losing any yourselves, I observe you to be affectionately lavish; but of the jewels of God, if any suggestions be made by charity touching the distribution of *them*, you are apt, in your wisdom, to make answer like the wise virgins, "Not so, lest there be not enough for us and you."

46. Now, my fair friends, doubtless, if the Tabernacle were to be erected again, in the middle of the Park, you would all be eager to stitch camels' hair for it;—some, to make presents of sealskins to it; and, perhaps, not a few fetch your jewel-cases, offering their contents to the selection of Bezaleel and Aholiab.

But that cannot be, now, with so Crystal-Palace-like entertainment to you. The tabernacle of God is now with men;—*in* men, and women, and sucklings also; which temple ye are, ye and your Christian sisters; of whom the poorest, here in London, are a very undecorated shrine indeed. *They* are the Tabernacle, fair friends, which you have got leave, and charge, to adorn. Not, in anywise, those charming churches and altars which you wreathe with garlands for God's sake, and the eloquent clergyman's. You are quite wrong, and barbarous in language, when you call *them* 'Churches' at all. They are only Synagogues;—the very same of which Christ spoke, with eternal meaning, as the places that hypocrites would love to be seen in. Here, in St. Giles's, and the East, sister to that in St. George's, and the West, is the Church! raggedly enough curtained, surely! Let those arches and pillars of Mr. Scott's alone, young ladies: it is *you* whom God likes to see well decorated, not them. Keep your roses for your hair—your embroidery for your petticoats. You are yourselves the Church, dears; and see that you be finally adorned, as women professing godliness, with the precious stones of good works, which may be quite briefly defined, for the present, as decorating the entire Tabernacle; and clothing your poor sisters, with yourselves. Put roses also in *their* hair, put precious stones also on *their* breasts; see that they also are clothed in your purple and scarlet, with other delights; that they also learn to read the gilded heraldry of the sky; and, upon the earth, be taught, not only the labours of it, but the loveliness. For them, also, let the hereditary jewel recall their father's pride, their mother's beauty: so shall your days, and theirs, be long in the sweet and sacred land which the Lord your God has given you: so, truly, shall THE GOLD OF THAT LAND BE GOOD, AND THERE, ALSO, THE CRYSTAL, AND THE ONYX STONE.

## CHAPTER VIII.

### THE ALPHABET.

(*Chapter written to introduce the preceding Lecture; but transposed, that the Lecture might not be divided between two numbers.*)

1. SINCE the last sentence of the preceding number of 'Deucalion' was written, I have been compelled, in preparing for the arrangement of my Sheffield museum, to look with nicety into the present relations of theory to knowledge in geological science; and find, to my no small consternation, that the assertions which I had supposed beyond dispute, made by the geologists of forty years back, respecting the igneous origin of the main crystalline masses of the primary rocks, are now all brought again into question; and that the investigations of many of the most intelligent observers render many former theories, in their generality, more than doubtful. My own studies of rock structure, with reference to landscape, have led me, also, to see the necessity of retreating to and securing the very bases of knowledge in this infinitely difficult science: and I am resolved, therefore,

at once to make the series of 'Deucalion' an absolutely trustworthy foundation for the geological teaching in St. George's schools; by first sifting what is really known from what is supposed; and then, out of things known, sifting what may be usefully taught to young people, from the perplexed vanity of prematurely systematic science.

2. I propose, also, in the St. George's Museum at Sheffield, and in any provincial museums hereafter connected with it, to allow space for two arrangements of inorganic substances; one for mineralogists, properly so called, and the general public; the other for chemists, and advanced students in physical science. The mineralogical collection will be fully described and explained in its catalogue, so that very young people may begin their study of it without difficulty, and so chosen and arranged as to be comprehensible by persons who have not the time to make themselves masters of the science of chemistry, but who may desire some accurate acquaintance with the aspect of the principal minerals which compose the world. And I trust, as I said in the preceding lecture, that the day is near when the knowledge of the native forms and aspects of precious stones will be made a necessary part of a lady's education; and knowledge of the nature of the soils, and the building stones, of his native country, a necessary part of a gentleman's.

3. The arrangement of the chemical collection I shall leave to any good chemist who will undertake it: I sup-

pose that now adopted by Mr. Maskelyne for the mineral collection in the British Museum may be considered as permanently authoritative.

But the mineralogical collection I shall arrange myself, as aforesaid, in the manner which I think likely to be clearest for simple persons; omitting many of the rarer elements altogether, in the trust that they will be sufficiently illustrated by the chemical series; and placing the substances most commonly seen in the earth beneath our feet, in an order rather addressed to the convenience of memory than to the symmetries of classification.

4. In the outset, therefore, I shall divide our entire collection into twenty groups, illustrated each by a separately bound portion of catalogue.

These twenty groups will illustrate the native states, and ordinary combinations, of nine solid oxides, one gaseous element (fluorine), and ten solid elements, placed in the following order :—

1. Silica.
2. Oxide of Titanium.
3. Oxide of Iron.
4. Alumina.
5. Potassa.
6. Soda.
7. Magnesia.
8. Calcium.
9. Glucina.
10. Fluorine.

11. Carbon.
12. Sulphur.
13. Phosphorus.
14. Tellurium.
15. Uranium.
16. Tin.
17. Lead.
18. Copper.
19. Silver.
20. Gold.

5. A few words will show the objects proposed by this limited arrangement. The three first oxides are placed in one group, on account of the natural fellowship and constant association of their crystals.

Added to these, the next group of the alkaline earths will constitute one easily memorable group of nine oxides, out of which, broadly and practically, the solid globe of the earth is made, containing in the cracks, rents, or volcanic pits of it, the remaining eleven substances, variously prepared for man's use, torment, or temptation.

6. I put fluorine by itself, on account of its notable importance in natural mineralogy, and especially in that of Cornwall, Derbyshire, and Cumberland: what I have to say of chlorine and iodine will be arranged under the same head; then the triple group of anomalous substances created for ministry by fire, and the seven-fold group of the great metals, complete the list of substances

which must be generally known to the pupils in St. George's schools. The phosphates, sulphates, and carbonates of the earths, will be given with the earths; and those of the metals, under the metals. The carburets, sulphurets, and phosphurets, * under carbon, sulphur, and phosphorus. Under glucina, given representatively, on account of its importance in the emerald, will be given what specimens may be desirable of the minor or auxiliary earths—baryta, strontia, etc.; and under tellurium and uranium, the auxiliary metals—platinum, columbium, etc., naming them thus together, under those themselves named from Tellus and Uranus. With uranium I shall place the cupreous micas, for their similarity of aspect.

7. The minerals referred to each of these twenty groups will be further divided, under separate letters, into such minor classes as may be convenient, not exceeding twenty: the letters being initial, if possible, of the name of the class; but the letters I and J omitted, that they may not be confused with numerals; and any letter of important sound in the mineral's name substituted for these, or for any other that would come twice over. Then any number of specimens may be catalogued under each letter.

For instance, the siliceous minerals which are the subject of study in the following lecture will be lettered thus:—

---

* I reject the modern term 'sulphide' unhesitatingly. It is as barbarous as 'carbide.'

A. Agate.
C. Carnelian.
H. Hyalite.
L. Chalcedony.
M. Amethyst.
O. Opal.
Q. Quartz.
S. Jasper.

In which list, M is used that we may not have A repeated, and will yet be sufficiently characteristic of Amethyst; and L, to avoid the repetition of C, may stand for Chalcedony; while S, being important in the sound of Jasper, will serve instead of excluded J, or pre-engaged A.

The complete label, then, on any (principally) siliceous mineral will be in such form as these following:—

1 A 1, meaning Silica, Agate, No. 1.
1 L 40, " Silica, Chalcedony, No. 40.
1 Q 520, " Silica, Quartz, No. 520.

8. In many of the classes, as in this first one of Silica, we shall not need all our twenty letters; but there will be a letter A to every class, which will contain the examples that explain the relation and connection of the rest. It happens that in Silica, the agates exactly serve this purpose; and therefore may have A for their proper initial letter. But in the case of other minerals, the letter A will not be the initial of the mineral's name, but the

indication of its character, as explanatory of the succeeding series.

Thus the specimen of gold, referred to as 20 A 1 in the preceding lecture, is the first of the series exhibiting the general method of the occurrence of native gold in the rocks containing it; and the complete series in the catalogue will be—

A. Native Gold, in various geological formations.
B. Branched Gold.
C. Crystalline Gold.
D. Dispersed Gold.
G. Granulate Gold.
K. Knitted Gold.
L. Leaf Gold.
M. Mossy Gold.
R. Rolled Gold.

**9.** It may be at once stated that I shall always retain the word 'branched' for minerals taking either of the forms now called 'arborescent' or 'dendritic.' The advance of education must soon make all students feel the absurdity of using the epithet 'tree-like' in Latin, with a different meaning from the epithet 'tree-like' in Greek. My general word 'branched' will include both the so-called 'arborescent' forms (meaning those branched in straight crystals), and the so-called 'dendritic' (branched like the manganese or oxide in Mocha stones;)

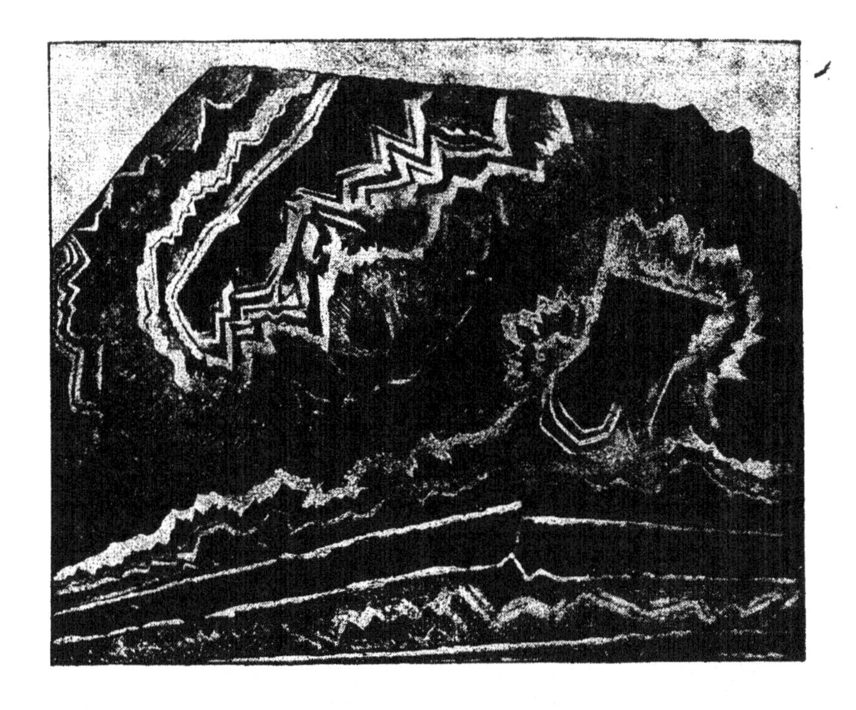

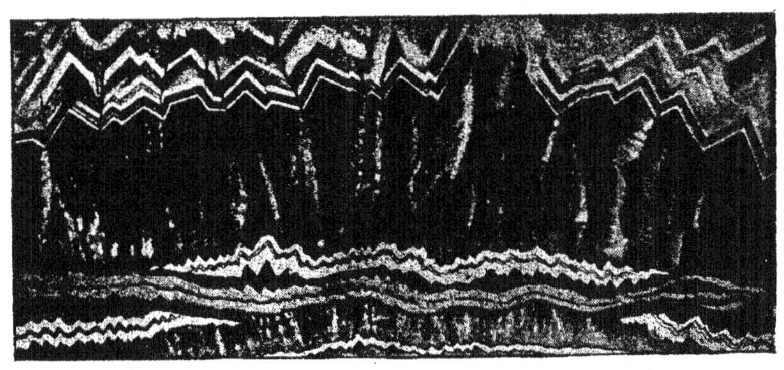

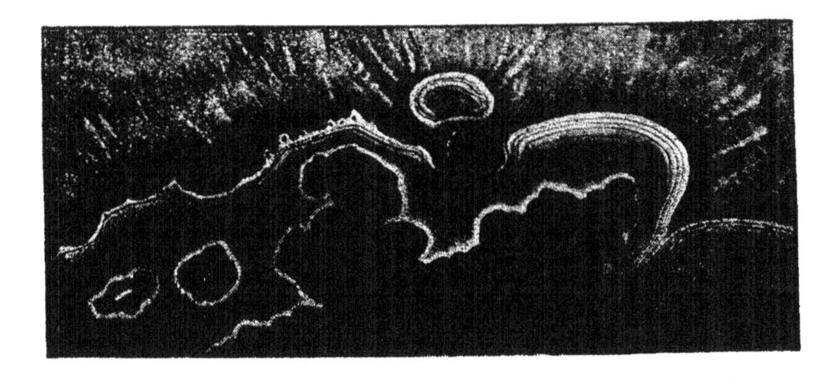

but with most accurate explanation of the difference; while the term 'spun' will be reserved for the variously thread-like forms, inaccurately now called dendritic, assumed characteristically by native silver and copper.

Of course, thread, branch, leaf, and grain, are all in most cases crystalline, no less definitely than larger crystals; but all my epithets are for practical service, not scientific definition; and I mean by 'crystalline gold' a specimen which distinctly shows octohedric or other specific form; and by 'branched gold' a specimen in which such crystalline forms are either so indistinct or so minute as to be apparently united into groups resembling branches of trees.

10. Every one of the specimens will be chosen for some specialty of character; and the points characteristic of it described in the catalogue; and whatever questions respecting its structure are yet unsolved, and significant, will be submitted in succession, noted each by a Greek letter, so that any given question may be at once referred to. Thus, for instance: question *a* in example 20 G 1 will be the relation of the subdivided or granular condition of crystalline gold to porous states of the quartz matrix. As the average length of description required by any single specimen, chosen on such principle, ought to be at least half a page of my usual type, the distribution of the catalogue into volumes will not seem unnecessary; especially as in due course of time, I hope that each volume will consist of two parts, the first contain-

ing questions submitted, and the second, solutions received.

The geological series will be distinguished by two letters instead of one, the first indicating the principal locality of the formation, or at least that whence it was first named. And I shall distinguish *all* formations by their localities—"M. L., Malham limestone"; "S. S., Skiddaw slate"; etc.,—leaving the geologists to assign systematic or chronological names as they like. What is pliocene to-day may be pleistocene to-morrow; and what is triassic in Mr. A.'s system, tesserassic in Mr. B.'s; but Turin gravels and Warwick sands remain where they used to be, for all that.

These particulars being understood, the lecture which I gave this spring on the general relations of precious minerals to human interests, may most properly introduce us to our detailed and progressive labour; and two paragraphs of it, incidentally touching upon methods of public instruction, may fitly end the present chapter.

11. In all museums intended for popular teaching, there are two great evils to be avoided. The first is, superabundance; the second, disorder. The first is having too much of everything. You will find in your own work that the less you have to look at, the better you attend. You can no more see twenty things worth seeing in an hour, than you can read twenty books worth reading in a day. Give little, but that little good and beautiful, and explain it thoroughly. For instance, here in crystal,

you may have literally a thousand specimens, every one with something new in it to a mineralogist; but what is the use of that to a man who has only a quarter of an hour to spare in a week? Here are four pieces—showing it in perfect purity,—with the substances which it is fondest of working with, woven by it into tissues as fine as Penelope's; and one crystal of it stainless, with the favourite shape it has here in Europe—the so-called 'flute-beak' of Dauphiné,—let a man once understand that crystal, and study the polish of this plane surface, given to it by its own pure growth, and the word 'crystal' will become a miracle to him, and a treasure in his heart for evermore.

12. Not too much, is the first law; not in disorder, is the second. Any order will do, if it is fixed and intelligible: no system is of use that is disturbed by additions, or difficult to follow; above all, let all things, for popular use, be *beautifully* exhibited. In our own houses, we may have our drawers and bookcases as rough as we please; but to teach our people rightly, we must make it a true joy to them to see the pretty things we have to show: and we must let them feel that, although, by poverty, they may be compelled to the pain of labour, they need not, by poverty, be debarred from the felicity and the brightness of rest; nor see the work of great artists, or of the great powers of nature, disgraced by commonness and vileness in the manner of setting them forth. Stateliness, splendour, and order are above all

things needful in places dedicated to the highest labours of thought: what we willingly concede to the Graces of Society, we must reverently offer to the Muses of Seclusion; and out of the millions spent annually to give attractiveness to folly, may spare at least what is necessary to give honour to Instruction.

# CHAPTER IX.

## FIRE AND WATER.

1. In examining any mineral, I wish my pupils first to be able to ascertain easily what it is; then to be accurately informed of what is *known* respecting the processes of its formation; lastly, to examine, with such precision as their time or instruments may permit, the effects of such formation on the substance. Thus, from almost any piece of rock, in Derbyshire, over which spring water has trickled or dashed for any length of time, they may break with a light blow a piece of brown incrustation, which, with little experience, they may ascertain to be carbonate of lime;—of which they may authoritatively be told that it was formed by slow deposition from the dripping water;—and in which, with little strain of sight, they may observe structural lines, vertical to the surface, which present many analogies with those which may be seen in coats of semi-crystalline quartz, or reniform chalcedony.

2. The more accurate the description they can give of the aspect of the stone, and the more authoritative and sifted the account they can render of the circumstances of its origin, the greater shall I consider their progress, and the more hopeful their scientific disposition.

But I absolutely forbid their proceeding to draw any logical inferences from what they know of stalagmite, to what they don't know of chalcedony. They are not to indulge either their reason or their imagination in the feeblest flight beyond the verge of actual experience; and they are to quench, as demoniacal temptation, any dispositiou they find in themselves to suppose that, because stalagmite and chalcedony both show lines of structure vertical to reniform surface, both have been deposited in a similar manner from a current solution. They are to address themselves to the investigation of the chalcedony precisely as if no stalagmite were in existence,—to inquire first what it is; secondly, when and how it is *known to be* formed; and, thirdly, what structure is discernible in it, —leaving to the close of their lives, and of other people's, the collection, from evidence thus securely accumulated, of such general conclusions as may then, without dispute, and without loss of time through prejudice in error, manifest themselves, not as 'theories,' but as demonstrable laws

When, however, for the secure instruction of my thus restrained and patient pupils, I look, myself, for what is actually told me by eye-witnesses, of the formation of mineral bodies, I find the sources of information so few, the facts so scanty, and the connecting paste, or diluvial detritus, of past guesses, so cumbrously delaying the operation of rational diamond-washing, that I am fain, as the shortest way, to set such of my friends as are minded to help me, to begin again at the very beginning; and *re*assert, for the general good, what their eyes can now see, in what their hands can now handle.

3. And as we have begun with a rolled flint, it seems by special guidance of Fors that the friend who has already first contributed to the art-wealth of the Sheffield Museum, Mr. Henry Willett, is willing also to be the first contributor to its scientific treasuries of fact; and has set himself zealously to collect for us the phenomena observable in the chalk and flint of his neighbourhood.

Of which kindly industry, the following trustworthy notes have been already the result, which, (whether the like observations have been made before or not being quite immaterial to the matter in hand,) are assuredly themselves original and secure: not mere traditional gossip. Before giving them, however, I will briefly mark their relations to the entire subject of the structure of siliceous minerals.

4. There are a certain number of rocks in the world, which have been seen by human eyes, flowing, white-hot, and watched by human eyes as they cool down. The structure of these rocks is therefore absolutely known tc have had something to do with fire.

There are a certain number of other rocks in the world which have been seen by human eyes in a state of wet sand or mud, and which have been watched, as they dried, into substances more or less resembling stone. The structure of these rocks is therefore known to have had something to do with water.

Between these two materials, whose nature is avouched by testimony, there occur an indefinite number of rocks, which no human eyes have ever seen, either hot or muddy; but which nevertheless show curious analogies to the ascertainably cooled substances on the one side, and to the ascertainably dried substances on the other. Respecting these medial formations, geologists have disputed in my ears during the half-century of my audient life; (and had been disputing for about a century before I was born,) without having yet arrived at any conclusion whatever; the book now held to be the principal authority on the subject, entirely contradicting, as aforesaid, the conclusions which, until very lately, the geological world, if it had not accepted as incontrovertible, at least asserted as positive.

5. In the said book, however,—Gustaf Bischof's Chemical Geology, —there are, at last, collected a large number of important and secure facts, bearing on mineral formation: and principles of microscopic investigation have been established by Mr. Sorby, some years ago, which have, I doubt not, laid the foundation, at last, of the sound knowledge of the conditions under which crystals are formed. Applying Mr. Sorby's method, with steady industry, to the rocks of Cumberland, Mr. Clifton Ward has, so far as I can judge, placed the nature of *these*, at least, within the range of secure investigation. Mr.

Ward's kindness has induced him also to spare the time needful for the test of the primary phenomena of agatescent structure in a similar manner; and I am engraving the beautiful drawings he sent me, with extreme care, for our next number; to be published with a letter from him, containing, I suppose, the first serviceable description of agatescent structure yet extant.*

6. Hitherto, however, notwithstanding all that has been accomplished, nobody can tell us how a common flint is made. Nobody ever made one; nobody has ever seen one naturally coagulate, or naturally dissolve; nobody has ever watched their increase, detected their diminution, or explained the exact share which organic bodies have in their formation. The splendid labours of Mr. Bowerbank have made us acquainted with myriads of organic bodies which have provoked siliceous concretion, or become entangled in it: but the beautiful forms which these present have only increased the difficulty of determining the real crystalline modes of siliceous structure, unaffected by organic bodies.

7. Crystalline *modes*, I say, as distinguished from crystalline *laws*. It is of great importance to mineralogy that we should carefully distinguish between the laws or limits which determine the possible angles in the form of

* I must, however, refer the reader to the valuable summary of work hitherto done on this subject by Professor Rupert Jones. (Proceedings of Geologists' Association, Vol. IV., No. 7,) for examination of these questions of priority.

a mineral, and the modes, or measures, in which, according to its peculiar nature or circumstances, it conducts itself under these restrictions.

Thus both cuprite and fluor are under laws which enforce cubic or octohedric angles in their crystals; but cuprite can arrange its cubes in fibres finer than those of the softest silk, while fluor spar only under rare conditions distinctly elongates its approximate cube into a parallelopiped.

Again, the prismatic crystals of Wavellite arrange themselves invariably in spherical or reniform concretions; but the rhombohedral crystals of quartz and hematite do so only under particular conditions, the study of which becomes a quite distinct part of their lithology.

8. This stellar or radiant arrangement is one essential condition in the forms and phenomena of agate and chalcedony; and Mr. Clifton Ward has shown in the paper to which I have just referred, that it is exhibited under the microscope as a prevalent condition in their most translucent substance, and on the minutest scale.

Now all siliceous concretions, distinguishing themselves from the mass of the surrounding rocks, are to be arranged under two main classes; briefly memorable as knots and nuts; the latter, from their commonly oval form, have been usually described by mineralogists as, more specially, 'almonds.'

'Knots' are concretions of silica round some central point or involved substance, (often organic); such knots being usually harder and more solid in the centre than at the outside, and having their fibres of crystallization, if visible, shot outwards like the rays of a star, forming pyramidal crystals on the exterior of the knot.

9. 'Almonds' are concretions of silica formed in cavities of rocks, or, in some cases, probably by their own energy producing the cavities they enclose; the fibres of crystallization, if visible, being directed from the outside of the almond-shell towards its interior cavity.

10. These two precisely opposite conditions are severally represented best by a knot of sound black flint in chalk, and by a well-formed hollow agate in a volcanic rock.

I have placed in the Sheffield Museum a block of black flint, formed round a bit of Inoceramus shell; and an almond-shell of agate, about six times as big as a cocoa nut, which will satisfactorily illustrate these two states. But between the two, there are two others of distinctly gelatinous silica, and distinctly crystalline silica, filling pores, cavities, and veins, in rocks, by infiltration or secretion. And each of these states will be found passing through infinite gradations into some one of the three others, so that separate account has to be given of every step in the transitions before we can rightly understand the main types.

11. But at the base of the whole subject lies, first, the clear understanding of the way a knot of solid crystalline substance—say, a dodecahedral garnet—forms itself out of

a rock-paste, say greenstone trap, without admitting a hairs-breadth of interstice between the formed knot and enclosing paste; and, secondly, clear separation in our thoughts, of the bands or layers which are produced by crystalline segregation, from those produced by successively accumulating substance. But the method of increase of crystals themselves, in an apparently undisturbed solution, has never yet been accurately described; how much less the phenomena resulting from influx of various elements, and changes of temperature and pressure. The frontispiece to the third number of 'Deucalion' gives typical examples of banded structure resulting from pure crystalline action; and the three specimens, 1. A. 21, 22, and 23, at Sheffield, furnish parallel examples of extreme interest. But a particular form of banding in flint, first noticed and described by Mr. S. P. Woodward,* is of more interest than any other in the total obscurity of its origin; and in the extreme decision of the lines by which, in a plurality of specimens, the banded spaces are separated from the homogeneous ones, indicating the first approach to the conditions which produce, in more perfect materials, the forms of, so-called, 'brecciated' agates. Together with these, a certain number of flints are to be examined which present every appearance of having been violently fractured and re-cemented. Whether fractured by mechanical violence, by the expansive or decomponent

* 'Geological Magazine,' 1864, vol. i., p. 145, pl. vii. and viii.

forces of contained minerals, or by such slow contraction and re-gelation as must have taken place in most veins through masses of rock, we have to ascertain by the continuance of such work as my friend has here begun.

### Letter I. *—*Introductory.*

12. "I am beginning to be perplexed about the number of flints, containing problems and illustrations, and wondering to what extent my inquiries will be of auv use to you.

"I intended at first to collect only what was really beautiful in itself—'crystalline'! but how the subject widens, and how the arbitrary divisions do run into one another! What a paltry shifting thing our classification is! One is sometimes tempted to give it all up in disgust, and I have a shrewd suspicion that all scientific classification (except for mutual aid to students) is absurd and pedantic: (*a*) varieties, species, genera, classes, orders, have most of them more in common than of divergence,—'a forming spirit' everywhere, for use and beauty.

---

* I shall put my own notes on these and any future communications I may insert, in small print at the bottom of the pages; and with letter-references—*a*, *b*, etc.; but the notes of the authors themselves will be put at the end of their papers, in large print, and with number-references—1, 2, etc.

(*a*) All, at least, is imperfect; and most of it absurd in the attempt to be otherwise.

"It is (to me) impossible to separate purely mineral and chemical siliceous bodies in chalk, (*b*) from those which are partly formed by the silicate-collecting sponges, which seem to have given them their forms.

"Who is to say that the radiations and accretions of a crystal are not life, but that the same arrangements in a leaf or a tree are life?—that the clouds which float in their balanced changeableness are not as much guided and defined as the clouds of the chalcedony, or the lenses of the human eye which perceives them?

"I think the following facts are plain:

"1. The chalk bands do go through the flint.

"2. Fissures in flints are constantly repaired by fresh deposits of chalcedony and silex.

"3. Original sponge matter is preserved (*c*) and obliterated by siliceous deposit, in extent and degree varying infinitely, and apparently proportioned to the amount of iron present—*i. e.*, the iron preserves original form, unless when combined with sulphur enough to crystallize, when all the original structure disappears.

"4. Amygdaloids seem to be formed by a kind of independent or diverse arrangement of molecules, caused by slight admixture of foreign minerals."

---

(*b*) It may be doubtful if any such exist in chalk; but, if they exist, they will eventually be distinguishable.

(*c*) Q. The form or body of it only; is the matter itself ever preserved?

LETTER II.—*Memoranda made at Mantell's Quarry, Cuckfield, on the banding noticed in the beds and nodules of the siliceous calciferous sandstone there, 31st May*, 1876.

Nos. I. and II. Ovate, concentric, ferruginous bandings; the centre apparently (1) free from banding.

III. Bands arranged at acute angles. These bands are not caused by fracture, but apparently by the intersection, at an acute angle, of the original lines of deposit. (*d*)

IV. In this specimen the newly fractured surfaces show no bandings, but the weathered surface develops the banding.

V. Ditto—*i. e.* bands parallel; much more ferrugi nous, and consequently more friable when exposed to weathering.

May not something be learnt regarding the laws of banding in agates, flints, etc., from observing the arrangement of banding in rocks composed mainly of siliceous matter? (*e*)

May not some of the subtler influences which regulate the growth of trees in their lines of annual increase (mag-

---

Note 1, page 170.

(*d*) These angular concretions require the closest study; see the segments of spheres in the plate given in the last number

(*e*) More, I should say, from the agates, respecting the laws of band ing in rocks: see the plate to the present number. When we can explain the interruptions of the bands on such scale as this, we may begin to understand some of those in larger strata.

netic probably) have some effect in the arrangement of minerals in solution?—nay, even of the higher vital processes, such as the deposition of osseous matter in teeth and bones? (*f*)

Letter III.—*Memoranda respecting banded chalk.*

I. In the banded lines (ferruginous) noticed above and below the horizontal fissures beneath the cliff at the Hope Gap, Seaford, it is evident that these lines are not markings of original deposition, but are caused by successive infiltrations of water containing iron in solution. (*g*)

II. Concentric markings of the same nature are observable in places where—

*a.* Iron pyrites are decomposing, and the iron in solution is being successively infiltrated into the surrounding chalk rock.

*b.* From dropping of ferruginous springs through crevices on horizontal surfaces.

*c.* This is observable also on surfaces of tabular flint.

III. Very peculiar contorted bandings, (similar to the so-called contorted-rocks,) are observable in certain places, notably in the face of the chalk-pit on the east side of Goldstone Bottom. This chalk-pit, or quarry, is remarkable—

(*f*) Yes, certainly; but in such case, the teeth and bones act by mineral law; not the minerals by teeth and bone law.

(*g*) Questionable. Bands are almost always caused by concretion, or separation, not infiltration. However caused, the essential point, in the assertion of which this paper has so great value, is their distinction from strata.

1. *For the contorted bandings in the chalk rock which are not markings of original deposition, being quite independent of original stratification.* (*h*)

2. For the excessive shattering and fissuring observable.

3. For the fact that these cracks and fissures have been refilled with distinctive and varying substances, as with flint, clay, Websterite, and intermediate admixtures of these substances.

4. For veins of flint, formerly horizontal, which show visible signs of displacement by subsidence.

5. For the numerous fissures in these veins of tabular flint being stained by iron, which apparently aids in the further process of splitting up and of widening the minute crevices in the flint. The iron also appears to be infiltrated at varying depths into the body of unfractured flint.

*Qy.* Has not ordinary flint the power or property of absorbing ferruginous fluid?

LETTER IV.—*Memoranda respecting brecciate flint.*

"*June* 7, 1876.

"I hasten to report the result of my fresh inquiry respecting the specimen I first sent to you as 'breccia,' but which you doubted.

(*h*) A most important point. It is a question with me whether the greater number of minor contortions in Alpine limestones may not have been produced in this manner. When once the bands are arranged by segregation, chemical agencies will soon produce mechanical separation, as of original beds.

"The site is the embouchure of the little tidal river Cuckmere, about two miles east of Seaford. I found a block at about the same spot (about three hundred yards east of the coastguard station, and about three quarters of the distance west of the river's mouth).

"The rocks are here covered with sand, or with a bed of the old valley alluvium, not yet removed by wave action. Travelling westward, the transported blocks of breccia gradually increase in size, (a pretty sure augury that they were derived from a western source). The whole coast is subject to a very rapid degradation and consequent encroachment of the sea, the average in some places being from twenty-five to thirty feet yearly. At a spot a hundred yards east of the coastguard station, blocks of one or two tons were visible. The denuded chalk rock is of chalk, seamed and fissured; the cliff of the same nature; but all the flints, and especially the tabular veins, are splintered and displaced to an unusual extent.

"Farther westward yet, the blocks of breccia weigh several tons, the cement being itself fissured, and in some places consisting of angular fragments stained with iron. From one mass I extracted a hollow circular flint split into four or five pieces, the fragments, although displaced, re-cemented in juxtaposition. (*i*)

---

(*i*) I am not prepared to admit, yet, that any of these phenomena are owing to violence. We shall see.

"At the Hope Gap, the whole cliff becomes a fractured mass, the fissures being refilled, sometimes with calcareous cement, sometimes with clay, and in other places being hollow.

"From the sides of an oblique fissure filled with clay I extracted two pieces of a nodular flint, separated from each other by a two-inch seam of clay: when replaced (the clay having been removed) the two fitted exactly. An examination of the rocks shows that the fissures, which run in all directions, are largest when *nearly horizontal*, dipping slightly seawards.

"The upper and lower portions of some of these horizontal fissures are banded with iron stains, evidently derived from iron-water percolating the seams.

"If I am right, therefore, the mystery seems to be explained thus: (*k*)—

"I. Rain water, charged with carbonic acid, falling on the hills behind, trickles past the grass and humus beneath, through the cracks in the chalk, dissolving the carbonate of lime into a soluble bi-carbonate. Falling downwards, it escapes seawards through the horizontal fissures, widening them by its solvent power.

"II. The weight of the superincumbent mass by slow,

---

(*k*) I think this statement of Mr. Willett's extremely valuable; and see no reason to doubt its truth, as an explanation of the subsidence of chalk and limestone in certain localities. I do not hitherto receive it as any explanation of fracture in flints. I believe Dover Cliffs might sink to Channel bottom without splitting a flint, unless bedded.

certain, irregular pressure, descends, maintaining the contact of surfaces, but still ever sinking at intervals, varied by the resisting forces of weight and pressure.

"III. This process is probably accelerated by the inflow and reflow of salt water at the ebb and flow of tide (into the fissures.)

"IV. At certain periods, probablv in the summer, (as soluble bi-carbonate of lime becomes less soluble as temperature increases,) a portion becomes redeposited as a hard semi-crystalline calcareous cement.

"V. This cement appears, in some instances, to be slightly siliceous, and may have a tendency, by the mutual attraction of siliceous matter, to form solid layers of tabular flint.

"VI. If these deductions be correct, it is probable that the great results involved in the sinking of limestone hills, and the consequent encroachment of the sea, may be traced (step by step) to the springs in valleys 'which run among the hills;' thence to the rain and dewdrops; higher up to the mists and clouds; and so onward, by solar heat, to the ocean, where at last again they find their rest."

LETTER V.—*Final Abstract.*

"*June* 13, 1876.

"In addition to the heat derived from summer and atmospheric changes, there will be a considerable amount of heat evolved from the friction produced between the sides of fissures when slipping and subsidence occur, and from the crushing down of flint supports when weight overcomes resistance.

"After heavy rainfall—

1. Fissures are filled.
2. Solution is rapid.
3. Hydraulic pressure increases.
4. Fissures are widened.

"After a period of dry weather—

1. Solution is diminished.
2. Hydraulic pressure relieved.
3. Subsidence and flint-crushing commence, or progress more rapidly.
4. Heat is evolved.
5. Carbonic acid discharged.
6. Semi-crystalline carbonate of lime is deposited around.
   *a.* Fragments of crushed flint, (at rest at intermitting intervals between motion of rocks).
   *b.* Angular fragments of original chalk rock.
   *c.* Angular fractured pieces of old cement.

"I have a dawning suspicion that siliceous deposits (as chalcedony, etc.) are made when the temperature falls, for reasons which I must postpone to a future paper."

---

(1) Probably the same arrangement exists (concentric), but has not been made visible because the iron has not been oxydized.

# CHAPTER X.

## 'THIRTY YEARS SINCE.'

VILLAGE OF SIMPLON, *2d September*, 1876.

1. I AM writing in the little one-windowed room opening from the salle-à-manger of the Hotel de la Poste; but under some little disadvantage, being disturbed partly by the invocation, as it might be fancied, of calamity on the heads of nations, by the howling of a frantic wind from the Col; and partly by the merry clattering of the knives and forks of a hungry party in the salon doing their best to breakfast adequately, while the diligence changes horses.

In that same room,—a little earlier in the year,—two-and-thirty years ago, my father and mother and I were sitting at one end of the long table in the evening: and at the other end of it, a quiet, somewhat severe-looking, and pale, English (as we supposed) traveller, with his wife; she, and my mother, working; her husband carefully completely some mountain outlines in his sketch-book.

2. Those days are become very dim to me; and I forget which of the groups spoke first. My father and

mother were always as shy as children; and our busy fellow-traveller seemed to us taciturn, slightly inaccessible, and even Alpestre, and, as it were, hewn out of mountain flint, in his serene labour.

Whether some harmony of Scottish accent struck my father's ear, or the pride he took in his son's accomplishments prevailed over his own shyness, I think we first ventured word across the table, with view of informing the grave draughtsman that *we* also could draw. Whereupon my own sketch-book was brought out, the pale traveller politely permissive. My good father and mother had stopped at the Simplon for me, (and now, feeling miserable myself in the thin air, I know what it cost them,) because I wanted to climb the high point immediately west of the Col, thinking thence to get a perspective of the chain joining the Fletschhorn to the Monte Rosa. I had been drawing there the best part of the afternoon, and had brought down with me careful studies of the Fletschhorn itself, and of a great pyramid far eastward, whose name I did not know, but, from its bearing, supposed it must be the Matterhorn, which I had then never seen.

3. I have since lost both these drawings; and if they were given away, in the old times when I despised the best I did, because it was not like Turner, and any friend has preserved them, I wish they might be returned to me; for they would be of value in Deucalion, and of greater value to myself; as having won for me, that evening, the

sympathy and help of James Forbes. For his eye grew keen, and his face attentive, as he examined the drawings; and he turned instantly to me as to a recognized fellow-workman,—though yet young, no less faithful than himself.

He heard kindly what I had to ask about the chain I had been drawing; only saying, with a slightly proud smile, of my peak supposed to be the Matterhorn,* "No, —and when once you have seen the Matterhorn, you will never take anything else for it!"

He told me as much as I was able to learn, at that time, of the structures of the chain, and some pleasant general talk followed; but I knew nothing of glaciers then, and he had his evening's work to finish. And I never saw him again.

I wonder if he sees me now, or guided my hand as I cut the leaves of M. Violet-le-Duc's 'Massif du Mont Blanc' this morning, till I came to page 58,—and stopped!

I must yet go back, for a little while, to those dead days.

4. Failing of Matterhorn on this side of the valley of the Rhone, I resolved to try for it from the other; and begged my father to wait yet a day for me at Brieg.

No one, then, had ever heard of the Bell Alp; and few English knew even of the Aletsch glacier. I laid my

---

* It was the Weisshorn.

plans from the top of the Simplon Col; and was up at four, next day; in a cloudless morning, climbing the little rock path which ascends directly to the left, after crossing the bridge over the Rhone, at Brieg; path which is quite as critical a little bit of walking as the Ponts of the Mer de Glace; and now, encumbered with the late fallen shatterings of a flake of gneiss of the shape of an artichoke leaf, and the size of the stern of an old ship of the line, which has rent itself away, and dashed down like a piece of the walls of Jericho, leaving exposed, underneath, the undulatory surfaces of pure rock, which, I am under a very strong impression, our young raw geologists take for real "muttoned" glacier tracks *

5. I took this path because I wanted first to climb the green wooded mass of the hill rising directly over the valley, so as to enfilade the entire profiles of the opposite chain, and length of the valley of the Rhone, from its brow.

By midday I had mastered it, and got up half as high again, on the barren ridge above it, commanding a little tarn; whence, in one panorama are seen the Simplon and Saas Alps on the south, with the Matterhorn closing the avenue of the valley of St. Nicolas; and the Aletsch Alps on the north, with all the lower reach of the Aletsch glacier. This panorama I drew carefully; and slightly

---

* I saw this wisely suggested in a recent number of the 'Alpine Journal.'

coloured afterwards, in such crude way as I was then able; and fortunately not having lost this, I place it in the Sheffield Museum, for a perfectly trustworthy witness to the extent of snow on the Breithorn, Fletschhorn, and Montagne de Saas, thirty years ago.

My drawing finished, I ran round and down obliquely to the Bell Alp, and so returned above the gorge of the Aletsch torrent—making some notes on it afterwards used in 'Modern Painters,' many and many such a day of foot and hand labour having been needed to build that book, in which my friends, nevertheless, I perceive, still regard nothing but what they are pleased to call its elegant language, and are entirely indifferent, with respect to that and all other books they read, whether the elegant language tells them truths or lies.

That book contains, however, (and to-day it is needful that I should not be ashamed in this confidence of boasting,) the first faithful drawings ever given of the Alps, not only in England, but in Europe; and the first definitions of the manner in which their forms have been developed out of their crystalline rocks.

6. 'Definitions' only, observe, and descriptions; but no 'explanations.' I knew, even at that time, far too much of the Alps to theorize on them; and having learned, in the thirty years since, a good deal more, with the only consequence of finding the facts more inexplicable to me than ever, laid M. Violet-le-Duc's book on the seat of the carriage the day before yesterday, among other

stores and preparations for passing the Simplon, contemplating on its open first page the splendid dash of its first sentence into space,—"La croute terrestre, refroidie au moment du plissement qui a formé le massif du Mont Blanc,"—with something of the same amazement, and same manner of the praise, which our French allies are reported to have rendered to our charge at Balaclava:—

"C'est magnifique;—mais ce n'est pas"—la geologie.

7. I soon had leisure enough to look farther, as the steaming horses dragged me up slowly round the first ledges of pines, under a drenching rain which left nothing but their nearest branches visible. Usually, their nearest branches, and the wreaths of white cloud braided among them, would have been all the books I cared to read; but both curiosity and vanity were piqued by the new utterances, prophetic, apparently, in claimed authority, on the matters timidly debated by me in old time.

I soon saw that the book manifested, in spite of so great false-confidence, powers of observation more true in their scope and grasp than can be traced in any writer on the Alps since De Saussure. But, alas, before we had got up to Berisal, I had found also more fallacies than I could count, in the author's first statements of physical law; and seen, too surely, that the poor Frenchman's keen natural faculty, and quite splendid zeal and industry, had all been wasted, through the wretched national vanity which made him interested in Mont Blanc only 'since it became a part of France,' and had thrown him totally into the clique of Agassiz and Desor, with results in which neither the clique, nor M. Violet, are likely, in the end, to find satisfaction.

8. Too sorrowfully weary of bearing with the provincial temper, and insolent errors, of this architectural restoration of the Gothic globe, I threw the book aside, and took up my Carey's Dante, which is always on the carriage seat, or in my pocket—not exactly for reading, but as an antidote to pestilent things and thoughts in general; and store, as it were, of mental quinine,—a few lines being usually enough to recover me out of any shivering marsh fever fit, brought on among foulness or stupidity.

It opened at a favourite old place, in the twenty-first canto of the Paradise, (marked with an M. long ago, when I was reading Dante through to glean his mountain descriptions):—

> "'Twixt either shore
> Of Italy, nor distant from thy land," etc.;

and I read on into the twenty-third canto, down to St. Benedict's

> "There, all things are, as they have ever been;
> Our ladder reaches even to that clime,
> Whither the patriarch Jacob saw it stretch
> Its topmost round, when it appeared to him
> With angels laden. But to mount it now
> None lifts his foot from earth; and hence my rule

Is left a profitless stain upon the leaves.
The walls, for abbey reared, turned into dens;
The cowls, to sacks choked up with musty meal.

* * * *

His convent, Peter founded without gold
Or silver; I, with prayers and fasting, mine;
And Francis his, in meek humility.
And if thou note the point whence each proceeds,
Then look what it hath erred to, thou shalt find
The white turned murky.
Jordan was turned back,
And a less wonder than the refluent sea
May, at God's pleasure, work amendment here."

9. I stopped at this, (holding myself a brother of the third order of St. Francis,) and began thinking how long it would take for any turn of tide by St. George's work, when a ray of light came gleaming in at the carriage window, and I saw, where the road turns into the high ravine of the glacier galleries, a little piece of the Breithorn snowfield beyond.

Somehow, I think, as fires never burn, so skies never clear, while they are watched; so I took up my Dante again, though scarcely caring to read more; and it opened, this time, not at an accustomed place at all, but at the "I come to aid thy wish," of St. Bernard, in the thirty-first canto. Not an accustomed place, because I always think it very unkind of Beatrice to leave him to St. Bernard; and seldom turn expressly to the passage: but it has chanced lately to become of more significance to me, and I read on eagerly, to the "So burned the peaceful oriflamme," when the increasing light became so strong that it awaked me, like a new morning; and I closed the book again, and looked out.

We had just got up to the glacier galleries, and the last films of rain were melting into a horizontal bar of blue sky which had opened behind the Bernese Alps.

I watched it for a minute or two through the alternate arch and pier of the glacier galleries, and then as we got on the open hill flank again, called to Bernardo * to stop.

10. Of all views of the great mountains that I know in Switzerland, I think this, of the southern side of the Bernese range from the Simplon, in general the most disappointing—for two reasons: the first, that the green mass of their foundation slopes so softly to the valley that it takes away half the look of their height; and the second, that the greater peaks are confused among the crags immediately above the Aletsch glacier, and cannot, in quite clear weather, be recognized as more distant, or more vast. But at this moment, both these disadvantages were totally conquered. The whole valley was full of absolutely impenetrable wreathed cloud, nearly all pure white, only the palest grey rounding the changeful domes of it; and beyond these domes of heavenly marble, the great Alps stood up against the blue,—not

* Bernardo Bergonza, of the Hotel d'Italie, Arona, in whom any friend of mine will find a glad charioteer; and they cannot anywhere find an abler or honester one.

wholly clear, but clasped and entwined with translucent folds of mist, traceable, but no more traceable, than the thinnest veil drawn over St. Catherine's or the Virgin's hair by Lippi or Luini; and rising as they were withdrawn from such investiture, into faint oriflammes, as if borne by an angel host far distant; the peaks themselves strewn with strange light, by snow fallen but that moment,—the glory shed upon them as the veil fled; —and intermittent waves of still gaining seas of light increasing upon them, as if on the first day of creation.

"À present, vous pouvez voir l'hotel sur le Bell Alp, bati par Monsieur Tyndall."

The voice was the voice of the driver of the supplementary pair of horses from Brieg, who, just dismissed by Bernardo, had been for some minutes considering how he could best recommend himself to me for an extra franc.

I not instantly appearing favourably stirred by this information, he went on with increased emphasis, "Monsieur le *professeur* Tyndall."

The poor fellow lost his bonnemain by it altogether—not out of any deliberate spite of mine; but because, at this second interruption, I looked at him, with an expression (as I suppose) so little calculated to encourage his hopes of my generosity that he gave the matter up in a moment, and turned away, with his horses, down the hill;—I partly not caring to be further disturbed, and being besides too slow—as I always am in cases where presence of mind is needful—in calling him back again.

11. For, indeed, the confusion into which he had thrown my thoughts was all the more perfect and diabolic, because it consisted mainly in the stirring up of every particle of personal vanity and mean spirit of contention which could be concentrated in one blot of pure black ink, to be dropped into the midst of my aerial vision.

Finding it totally impossible to look at the Alps any more, for the moment, I got out of the carriage, sent it on to the Simplon village; and began climbing, to recover my feelings and wits, among the mossy knolls above the convent.

They were drenched with the just past rain; glittering now in perfect sunshine, and themselves enriched by autumn into wreaths of responding gold.

The vast hospice stood desolate in the hollow behind them; the first time I had ever passed it with no welcome from either monk, or dog. Blank as the fields of snow above, stood now the useless walls; and for the first time, unredeemed by association; only the thin iron cross in the centre of the roof remaining to say that this had once been a house of Christian Hospitallers.

12. Desolate this, and dead the office of this,—for the present, it seems; and across the valley, instead, "l'hotel sur le Bell Alp, bati par Monsieur Tyndall," no nest of dreamy monks, but of philosophically peripatetic or perisaltatory 'puces des glaces.'

For, on the whole, that is indeed the dramatic aspect and relation of them to the glaciers; little jumping black things, who appear, under the photographic microscope, active on the ice-waves, or even inside of them;—giving to most of the great views of the Alps, in the windows at Geneva, a more or less animatedly punctuate and pulicarious character.

Such their dramatic and picturesque function, to any one with clear eyes; their intellectual function, however, being more important, and comparable rather to a symmetrical succession of dirt-bands,—each making the ice more invisible than the last; for indeed, here, in 1876, are published, with great care and expense, such a quantity of accumulated rubbish of past dejection, and moraine of finely triturated mistake, clogging together gigantic heaped blocks of far-travelled blunder,—as it takes away one's breath to approach the shadow of.

13. The first in magnitude, as in origin, of these long-sustained stupidities,—the pierre-à-Bot, or Frog-stone, par excellence, of the Neuchâtel clique,—is Charpentier's Dilatation Theory, revived by M. Violet, not now as a theory, but an assured principle!—without, however, naming Charpentier as the author of it; and of course without having read a word of Forbes's demolition of it. The essential work of Deucalion is construction, not demolition; but when an avalanche of old rubbish is shot in our way, I must, whether I would or no, clear it aside before I can go on. I suppose myself speaking to my

Sheffield men; and shall put so much as they need know of these logs upon the line, as briefly as possible, before them.

14. There are three theories extant, concerning glacier-motion, among the gentlemen who live at the intellectual 'Hotel des Neuchâtelois.' These are specifically known as the Sliding,—Dilatation,—and Regelation, theories.

When snow lies deep on a sloping roof, and is not supported below by any cornice or gutter, you know that when it thaws, and the sun has warmed it to a certain extent, the whole mass slides off into the street.

That is the way the scientific persons who hold the 'Sliding theory,' suppose glaciers to move. They assume, therefore, two things more; namely, first that all mountains are as smooth as house-roofs; and, secondly, that a piece of ice a mile long and three or four hundred feet deep will slide gently, though a piece a foot deep and a yard long slides fast,—in other words, that a paving-stone will slide fast on another paving-stone, but the Rossberg fall at the rate of eighteen inches a day.

There is another form of the sliding theory, which is that glaciers slide in little bits, one at a time; or, for example, that if you put a railway train on an incline, with loose fastening to the carriages, the first carriage will slide first, as far as it can go, and then stop; then the second start, and catch it up, and wait for the third; and so on, till when the last has come up, the first will start again.

Having once for all sufficiently explained the 'Sliding theory' to you, I shall not trouble myself any more in Deucalion about it.

15. The next theory is the Dilatation theory. The scientific persons who hold *that* theory suppose that whenever a shower of rain falls on a glacier, the said rain freezes inside of it; and that the glacier being thereby made bigger, stretches itself uniformly in one direction, and never in any other; also that, although it can only be thus expanded in cold and wet weather, such expansion is the reason that it always goes fastest in hot and dry weather.

There is another form of the Dilatation theory, which is that the glacier expands by freezing its own meltings.

16. Having thus sufficiently explained the Dilatation theory to you, I shall not trouble myself in Deucalion farther about *it;* noticing only, in bidding it goodbye, the curious want of power in scientific men, when once they get hold of a false notion, to perceive the commonest analogies implying its correction. One would have thought that, with their thermometer in their hand to measure congelation with, and the idea of expansion in their head, the analogy between the tube of the thermometer, and a glacier channel, and the ball of the thermometer, and a glacier reservoir, might, some sunshiny day, have climbed across the muddily-fissured glacier of their wits:—and all the quicker, that their much-studied Mer de Glace bears to the great reservoirs of ice above it pre-

cisely the relation of a very narrow tube to a very large ball. The vast 'instrument' seems actually to have been constructed by Nature, to show to the dullest of savants the difference between the steady current of flux through a channel of drainage, and the oscillatory vivacity of expansion which they constructed their own tubular apparatus to obtain!

17. The last popular theory concerning glaciers is the Regelation theory. The scientific persons who hold *that* theory, suppose that a glacier advances by breaking itself spontaneously into small pieces; and then spontaneously sticking the pieces together again;—that it becomes continually larger by a repetition of this operation, and that the enlargement (as assumed also by the gentlemen of the Dilatation party), can only take place downwards.

You may best conceive the gist of the Regelation theory by considering the parallel statement, which you may make to your scientific young people, that if they put a large piece of barleysugar on the staircase landing, it will walk downstairs by alternately cracking and mending itself.

I shall not trouble myself farther, in Deucalion, about the Regelation theory.

18. M. Violet-le-Duc, indeed, appears to have written his book without even having heard of it; but he makes most dextrous use of the two others, fighting, as it were, at once with sword and dagger; and making his glaciers move on the Sliding theory when the ground is steep, and

on the Dilatation theory when it is level. The woodcuts at pages 65, 66, in which a glacier is represented dilating itself up a number of hills and down again, and that at page 99, in which it defers a line of boulders, which by unexplained supernatural power have been deposited all across it, into moraines at its side, cannot but remain triumphant among monuments of scientific error,—bestowing on their author a kind of St. Simeon-Stylitic pre-eminence of immortality in the Paradise of Fools.

19. Why I stopped first at page 58 of this singular volume, I see there is no room to tell in this number of Deucalion; still less to note the interesting repetitions by M. Violet-le-Duc of the Tyndall-Agassiz demonstration that Forbes' assertion of the plasticity of ice in large pieces, is now untenable, by reason of the more recent discovery of its plasticity in little ones. I have just space, however, for a little woodcut from the 'Glaciers of the Alps,' (or 'Forms of Water,' I forget which, and it is no matter,) in final illustration of the Tyndall-Agassiz quality of wit.

20. Fig. 5, A, is Professor Tyndall's illustration of the effect of sunshine on a piece of glacier, originally of the form shown by the dotted line, and reduced by solar power on the south side to the beautifully delineated wave in the shape of a wedge.

It never occurred to the scientific author that the sunshine would melt some of the top, as well as of the side, of his parallelopiped; nor that, during the process, even on the shady side of it, some melting would take place in the summer air. The figure at B represents three stages of the diminution which would really take place, allowing for these other somewhat important conditions of the

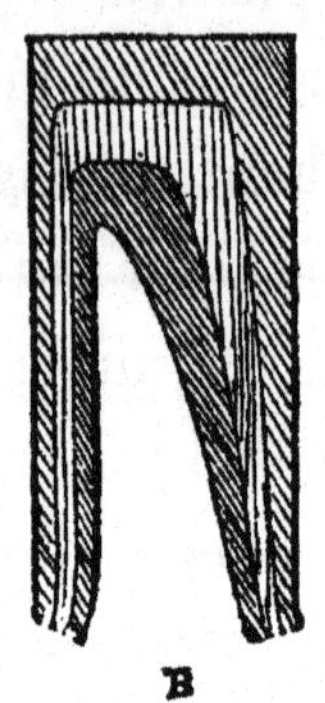


FIG. 5.

question; and it shows, what may farther interest the ordinary observer, how rectangular portions of ice, originally produced merely by fissure in its horizontal mass, may be gradually reduced into sharp, axe-edged ridges, having every appearance of splintery and vitreous fracture. In next Deucalion I hope to give at last some account of my experiments on gelatinous fracture, made in the delightful laboratory of my friend's kitchen, with the aid of her infinitely conceding, and patiently collaborating, cook.

# CHAPTER XI.

## OF SILICA IN LAVAS.

1. The rocks through whose vast range, as stated in the ninth chapter, our at first well-founded knowledge of their igneous origin gradually becomes dim, and fades into theory, may be logically divided into these four following groups.

I. True lavas. Substances which have been rapidly cooled from fusion into homogeneous masses, showing no clear traces of crystallization.

II. Basalts.* Rocks in which, without distinct separation of their elements, a disposition towards crystalline structure manifests itself.

III. Porphyries. Rocks in which one or more mineral elements separate themselves in crystalline form from a homogeneous paste.

IV. Granites. Rocks in which all their elements have taken crystalline form.

2. These, I say, are logical divisions, very easily tenable. But Nature laughs at logic, and in her infinite imagina-

---

* I use this word as on the whole the best for the vast class of rocks I wish to include; but without any reference to columnar desiccation. I consider, in this arrangement, only internal structure.

tion of rocks, defies all Kosmos, except the mighty one which we, her poor puppets, shall never discern. Our logic will help us but a little way;—so far, however, we will take its help.

3. And first, therefore, let us ask what questions imperatively need answer, concerning indisputable lavas, seen by living human eyes to flow incandescent out of the earth, and thereon to cool into ghastly slags.

On these I have practically burnt the soles of my boots, and in their hollows have practically roasted eggs; and in the lee of them, have been wellnigh choked with their stench; and can positively testify respecting them, that they were in many parts once fluid under power of fire, in a very fine and soft flux; and did congeal out of that state into ropy or cellular masses, variously tormented and kneaded by explosive gas; or pinched into tortuous tension, as by diabolic tongs; and are so finally left by the powers of Hell, to submit themselves to the powers of Heaven, in black or brown masses of adamantine sponge without water, and horrible honeycombs without honey, interlaid between drifted banks of earthy flood, poured down from merciless clouds whose rain was ashes.

The seas that now beat against these, have shores of black sand; the peasant, whose field is in these, ploughs with his foot, and the wind harrows.

4. Now of the outsides of these lava streams, and unaltered volcanic ashes, I know the look well enough; and could supply Sheffield with any quantity of characteristic

specimens, if their policy and trade had not already pretty nearly buried them, and great part of England besides, under such devil's ware of their own production. But of the *insides* of these lava streams, and of the recognized alterations of volcanic tufa, I know nothing. And, accordingly, I want authentic answer to these following questions, with illustrative specimens.

5. *a.* In lavas which have been historically hot to perfeet fusion, so as to be progressive, on steep slopes, in the manner of iron out of a furnace in its pig-furrows;—in such perfect lavas, I say—what kind of difference is there between the substance at the surface and at the extremest known depths, after cooling? It is evident that such lavas can only accumulate to great depths in infernal pools or lakes. Of such lakes, which are the deepest known? and of those known, where are the best sections? I want for Sheffield a series of specimens of any well-fused lava anywhere, showing the gradations of solidity or crystalline consolidation, from the outside to extreme depth.

*b.* On lavas which have not been historically hot, but of which there is no possible doubt that they were once fluent, (in the air,) to the above-stated degree, what changes are traceable, produced, irrespectively of atmospheric action, by lapse of time? What evidence is there that lavas, once cool to their centres, can sustain any farther crystalline change, or re-arrangement of mineral structure?

*c.* In lavas either historically or indisputably once fluent, what forms of silica are found? I limit myself at present to the investigation of volcanic *silica:* other geologists will in time take up other minerals; but I find silica enough, and more than enough, for my life, or at least for what may be left of it

Now I am myself rich in specimens of Hyalite, and Auvergne stellar and guttate chalcedonies; but I have no notion whatever how these, or the bitumen associated with them, have been developed; and I shall be most grateful for a clear account of their locality,—possible or probable mode of production in that locality,—and microscopic structure. Of pure quartz, of opal, or of agate, I have no specimen connected with what I should call a truly 'living' lava; one, that is to say, which has simply cooled down to its existing form from the fluid state; but I have sent to the Sheffield Museum a piece of Hyalite, on a living lava, so much like a living wasp's nest, and so incredible for a lava at all to the general observer, that I want forthwith some help from my mineralogical friends, in giving account of it.

6. And here I must, for a paragraph or two, pass from definition of flinty and molten minerals, to the more difficult definition of flinty and molten hearts; in order to explain why the Hyalite which I have just sent to the men of Sheffield, for their first type of volcanic silica,* is not at all the best Hyalite in my collection.

* I give the description of these seven pieces of Hyalite at Sheffield,

This is because I practically find a certain quantity of selfishness necessary to live by; and having no manner of saintly nature in me, but only that of ordinary men,—(which makes me all the hotter in temper when I can't get ordinary men either to see what I know they can see if they look, or do what I know they can do if they like,)—I get sometimes weary of giving things away, letting my drawers get into disorder, and losing the powers of observation and thought which are connected with the complacency of possession, and the pleasantness of order. Whereupon I have resolved to bring my own collection within narrow limits; but to constitute it resolutely and irrevocably of chosen and curious pieces, for my own pleasure; trusting that they may be afterwards cared for by some of the persons who knew me, when I myself am troubled with care no more.*

7. This piece of Hyalite, however, just sent to Sheffield, though not my best, is the most curiously *definite* example I ever saw. It is on a bit of brown lava, which looks, as aforesaid, a little way off, exactly like a piece

---

in Deucalion, because their description is necessary to explain certain general principles of arrangement and nomenclature.

* By the way, this selfish collection is to be primarily of stones that will *wash*. Of petty troubles, none are more fretting than the effect of dust on minerals that can neither be washed nor brushed. Hence, my specialty of liking for silica, felspar, and the granitic or gneissic rocks.

of a wasp's nest: seen closer, the cells are not hexagonal, but just like a cast of a spoonful of pease; the spherical hollows having this of notable in them, that they are only as close to each other as they can be, to *admit of their being perfectly round:* therefore, necessarily, with little spaces of solid stone between them. I have not the slightest notion how such a lava can be produced. It is like an oolite with the yolks of its eggs dropped out, and not in the least like a ductile substance churned into foam by expansive gas.

8. On this mysterious bit of gaseous wasp's nest, the Hyalite seems to have been dropped, like drops of glass from a melting glass rod. It seems to touch the lava just as little as it can; sticks at once on the edges of the cells, and laps over without running into, much less filling them. There is not any appearance, and I think no possibility, of exudation having taken place; the silica cannot but, I think, have been deposited; and it is stuck together just as if it had fallen in drops, which is what I mean by calling Hyalite characteristically 'guttate'; but it shows, nevertheless, a tendency to something like crystallization, in irregularities of surface like those of glacier ice, or the kind of old Venetian glass which is rough, and apparently of lumps coagulated. The fracture is splendidly vitreous,—the substance, mostly quite clear, but in parts white and opaque.

9. Now although no other specimen that I have yet seen is so manifestly guttate as this, all the hyalites I

know agree in approximate conditions; and associate themselves with forms of chalcedony which exactly resemble the droppings from a fine wax candle. Such heated waxen effluences, as they congeal, will be found thrown into flattened coats; and the chalcedonies in question on the *under* surface precisely resemble them; while on the *upper* they become more or less crystalline, and, in some specimens, form lustrous stellar crystals in the centre.

10. Now, observe, this chalcedony, *capable of crystallization*, differs wholly from chalcedony properly so called, which may indeed be *covered* with crystals, but itself remains consistently smooth in surface, as true Hyalite does, also.

Not to be teazed with too many classes, however, I shall arrange these peculiar chalcedonies with Hyalite· and, accordingly, I send next to the Sheffield Museum, to follow this first Hyalite, an example of the transition from Hyalite to dropped chalcedony, (I. H. 2,) being an Indian volcanic chalcedony, translucent, aggregated like Hyalite, and showing a *concave* fracture where a ball of it has been broken out.

11. Next, (I. H. 3,) pure dropped chalcedony. I do not like the word 'dropped' in this use,—so that, instead, I shall call this in future *wax* chalcedony; then (I. H. 4) the same form, with crystalline surface,—this I shall henceforward call *sugar* chalcedony; and, lastly, the ordinary stellar form of Auvergne, *star* chalcedony (I. H. 5).

These five examples are typical, and perfect in their kind; next to them (I. H. 6) I place a wax chalcedony formed on a porous rock, (volcanic ash?) which has at the surface of it small circular *concavities*, being also so irregularly coagulate throughout that it suggests no mode of deposition whatever, and is peculiar in this also, that it is thinner in the centre than at the edges, and that no vestige of its substance occurs in the pores of the rock it overlies.

Take a piece of porous broken brick, drop any tallowy composition over four or five inches square of its surface, to the depth of one-tenth of an inch; then drop more on the edges till you have a rampart round, the third of an inch thick; and you will have some likeness of this piece of stone: but how Nature held the composition in her fingers, or composed it to be held, I leave you to guess, for I cannot.

12. Next following, I place the most singular example of all (I. H. 7). The chalcedony in I. H. 6 is apparently dropped on the ashes, and of irregular thickness; it is difficult to understand *how* it was dropped, but once *get* Nature to hold the candle, and the thing is done.

But here, in I. H. 7, it is no longer apparently dropped, but apparently boiled! It rises like the bubbles of a strongly boiling liquid;—but not from a liquid mass; on the contrary, (except in three places, presently to be described,) it coats the volcanic ash in perfectly even thickness—a quarter of an inch, *and no more, nor less, everywhere*, over a space five inches square! and the ash, or lava, itself, instead of being porous throughout the mass, with the silica only on the surface, is filled with chalcedony in every cavity!

Now this specimen completes the transitional series from hyalite to perfect chalcedony; and with these seven specimens, in order, before us, we can define some things, and question of others, with great precision.

13. First, observe that all the first six pieces agree in two conditions,—*varying*, and *coagulated*, thickness of the deposit. But the seventh has the remarkable character of *equal*, and therefore probably crystalline, deposition everywhere.

Secondly. In the first six specimens, though the coagulations are more or less rounded, none of them are regularly spherical. But in the seventh, though the larger bubbles (so to call them) are subdivided into many small ones, every uninterrupted piece of the surface is *a portion of a sphere*, as in true bubbles.

Thirdly. The sugar chalcedony, I. H. 4, and stellar chalcedony, I. H. 5, show perfect power of assuming, under favourable conditions, prismatic crystalline form. But there is no trace of such tendency in the first three, or last two, of the seven examples. Nor has there ever, so far as I know, been found prismatic true hyalite, or prismatic true chalcedony.

Therefore we have here essentially three different minerals, passing into each other, it is true; but, at a certain point, changing their natures definitely, so that *hyalite, becoming wax chalcedony, gains* the power of prismatic crystallization; and *wax chalcedony, becoming true chalcedony, loses* it again!

And now I must pause, to explain rightly this term 'prismatic,' and others which are now in use, or which are to be used, in St. George's schools, in describing crystallization.

14. A prism, (the *sawn* thing,) in Newton's use of the word, is a triangular pillar with flat top and bottom. Putting two or more of these together, we can make pillars of any number of plane sides, in any regular or irregular shape. Crystals, therefore, which are columnar, and thick enough to be distinctly seen, are called 'prismatic.'

2. But crystals which are columnar, and so delicate that they look like needles, are called 'acicular,' from acus, a needle.
3. When such crystals become so fine that they look like hair or down, and lie in confused directions, the mineral composed of them is called 'plumose.'
4. And when they adhere together closely by their sides, the mineral is called 'fibrous.'
5. When a crystal is flattened by the extension of two of its planes, so as to look like a board, it is called 'tabular'; but people don't call it a 'tabula.'
6. But when such a board becomes very thin, it *is* called a 'lamina,' and the mineral composed of many such plates, laminated.

7. When laminæ are so thin that, joining with others equally so, they form fine leaves, the mineral is 'foliate.'
8. And when these leaves are capable of perpetual subdivision, the mineral is 'micaceous.'

15. Now, so far as I know their works, mineralogists hitherto have never attempted to show cause why some minerals rejoice in longitude, others in latitude, and others in platitude. They indicate to their own satisfaction,—that is to say, in a manner totally incomprehensible by the public,—all the modes of expatiation possible to the mineral, by cardinal points on a sphere: but why a crystal of ruby likes to be short and fat, and a crystal of rutile, long and lean; why amianth should bind itself into bundles of threads, cuprite weave itself into tissues, and silver braid itself into nests,—the use, in fact, that any mineral makes of its opportunities, and the cultivation which it gives to its faculties,—of all this, my mineralogical authorities tell me nothing. Industry, indeed, is theirs to a quite infinite degree, in pounding, decocting, weighing, measuring, but they have remained just as unconscious as vivisecting physicians that all this was only the anatomy of dust,—not its history.

But here at last, in Cumberland, I find a friend, Mr. Clifton Ward, able and willing to begin some true history of mineral substance, and far advanced already in preliminary discovery; and in answer to my request for help, taking up this first hyalitic problem, he has sent me the drawings—engraved, I regret to say, with little justice to their delicacy,*—in Plate V.

16. This plate represents, in Figure 1, the varieties of structure in an inch vertical section of a lake-agate; and in Figures 2, 3, 4, and 5, still farther magnified portions of the layers so numbered in Figure 1.

Figures 6 to 9 represent the structure and effect of polarized light in a lake-agate of more distinctly crystalline structure; and Figures 10 to 13, the orbicular concretions of volcanic Indian chalcedony. But before entering farther on the description of these definitely concretionary bands, I think it will be desirable to take note of some facts regarding the larger bands of our Westmoreland mountains, which become to me, the more I climb them, mysterious to a point scarcely tolerable; and only the more so, in consequence of their recent more accurate survey.

17. Leaving their pebbles, therefore, for a little while, I will ask my readers to think over some of the conditions of their crags and pools, explained as best I could, in the following lecture, to the Literary and Scientific Society of the town of Kendal. For indeed, beneath the evermore blessed Kendal-green of their sweet meadows and moors, the secrets of hill-structure remain, for all the work spent on them, in colourless darkness; and indeed, "So dark, Hal, that thou could'st not see thine hand."

---

* But not by my fault, for I told the engraver to do his best; and took more trouble with the plate than with any of my own.

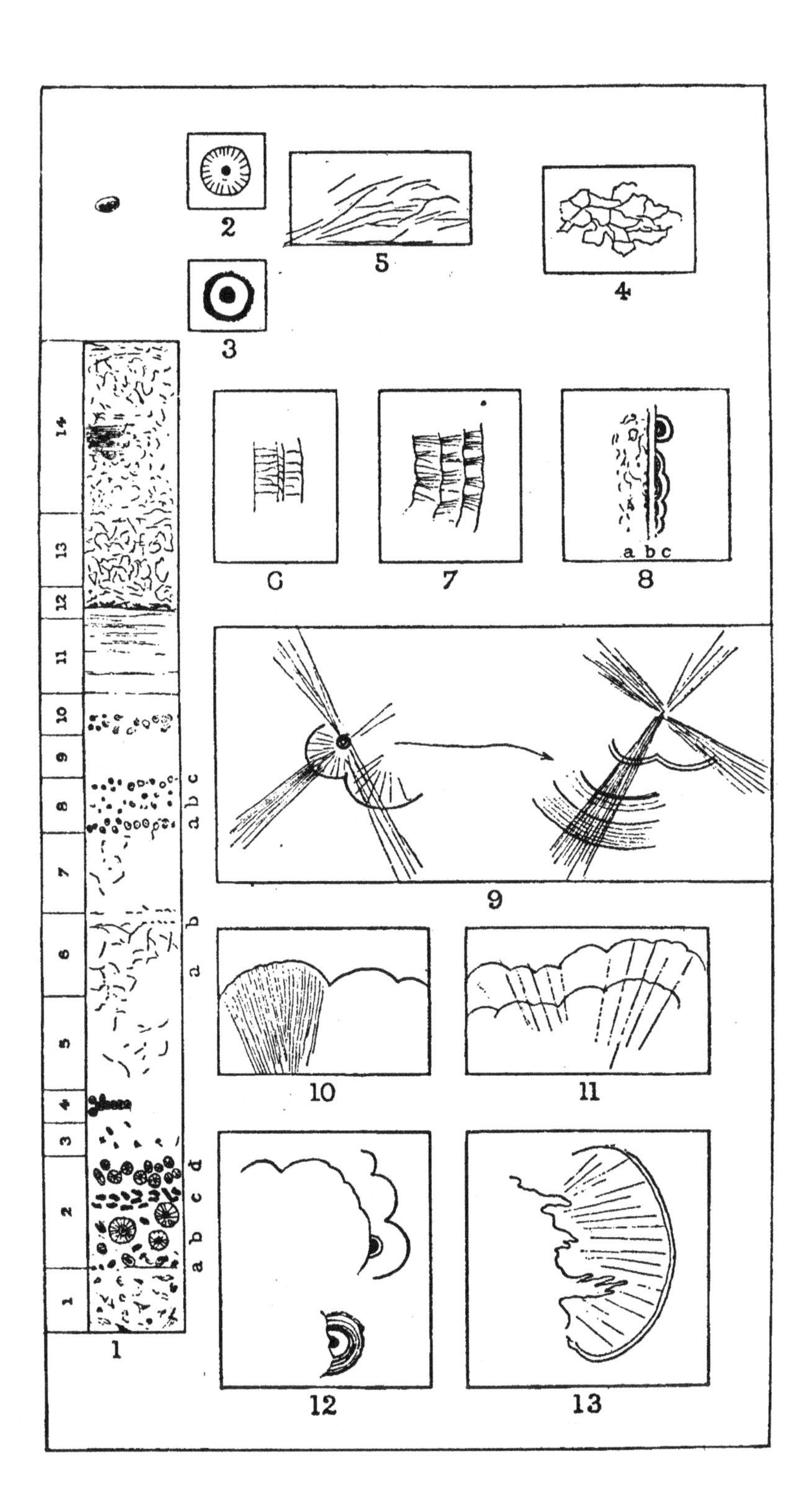
2
5
4
3
14
13
12
11
10
9
8
7
6
5
4
3
2
1
a b c
a b
a b c d
6
7
a b c
8
9
10
11
12
13
1

# CHAPTER XII.

## YEWDALE AND ITS STREAMLETS.

*Lecture delivered before the Members of the Literary and Scientific Institution, Kendal, 1st October,* 1877

1. I FEAR that some of my hearers may think an apology due to them for having brought, on the first occasion of my being honoured by their audience, a subject before them which they may suppose unconnected with my own special work, past or present. But the truth is, I knew mountains long before I knew pictures; and these mountains of yours, before any other mountains. From this town, of Kendal, I went out, a child, to the first joyful excursions among the Cumberland lakes, which formed my love of landscape and of painting: and now, being an old man, I find myself more and more glad to return—and pray you to-night to return with me—from shadows to the reality.

I do not, however, believe that one in a hundred of our youth, or of our educated classes, out of directly scientific circles, take any real interest in geology. And for my own part, I do not wonder,—for it seems to me that geology tells us nothing really interesting. It tells us much about a world that once was. But, for my part, a world that only was, is as little interesting as a world that only is to be. I no more care to hear of the forms of mountains that crumbled away a million of years ago to leave room for the town of Kendal, than of forms of mountains that some future day may swallow up the town of Kendal in the cracks of them. I am only interested—so ignoble and unspeculative is my disposition—in knowing how God made the Castle Hill of Kendal, for the Baron of it to build on, and how he brought the Kent through the dale of it, for its people and flocks to drink of.

2. And these things, if you think of them, you will find are precisely what the geologists cannot tell you. They never trouble themselves about matters so recent, or so visible; and while you may always obtain the most satisfactory information from them respecting the congelation of the whole globe out of gas, or the direction of it in space, there is really not one who can explain to you the making of a pebble, or the running of a rivulet.

May I, however, before pursuing my poor little inquiry into these trifling matters, congratulate those members of my audience who delight more in literature than science, on the possession, not only of dales in reality, but of dales in name. Consider, for an instant or two, how much is involved, how much indicated, by our possession in English of the six quite distinct words—vale, valley, dale, dell, glen, and dingle;—consider the gradations of character in scene, and fineness of observation in the inhabitants, implied by that sixfoil cluster of words · as compared to the simple 'thal' of the Germans, 'valle' of the Italians, and 'vallée' of the French, shortening into 'val' merely for ease of pronunciation, but having no variety of sense whatever; so that, supposing I want to translate, for the benefit of an Italian friend, Wordsworth's 'Reverie of Poor Susan,' and come to "Green pastures she views in the midst of the dale," and look for 'dale' in my Italian dictionary, I find "valle lunga e stretta tra poggi alti," and can only convey Mr. Wordsworth's meaning to my Italian listener by telling him that "la povera Susanna vede verdi prati, nel mezzo della valle lunga e stretta tra poggi alti"! It is worth while, both for geological and literary reasons, to trace the essential differences in the meaning and proper use of these words.

3. 'Vale' signifies a large extent of level land, surrounded by hills, or nearly so; as the Vale of the White Horse, or Vale of Severn. The level extent is necessary to the idea; while the next word, 'valley,' means a large hollow among hills, in which there is little level ground, or none. Next comes 'dale,' which signifies properly a tract of level land on the borders of a stream, continued for so great a distance as to make it a district of importance as a part of the inhabited country; as Ennerdale, Langdale, Liddesdale. 'Dell' is to dale, what valley is to vale; and implies that there is scarcely any level land beside the stream. 'Dingle' is such a recess or dell

clothed with wood;* and 'gleu' one varied with rocks. The term 'ravine,' a rent chasm among rocks, has its necessary parallel in other languages.

Our richness of expression in these particulars may be traced to the refinement of our country life, chiefly since the fifteenth century; and to the poetry founded on the ancient character of the Border peasantry; mingling agricultural with shepherd life in almost equal measure.

I am about to endeavour, then, to lay before you this evening the geological laws which have produced the 'dale,' properly so called, of which I take—for a sweet and near example—the green piece of meadow land through which flows, into Coniston Water, the brook that chiefly feeds it.

4. And now, before going farther, let me at once vindicate myself from the blame of not doing full justice to the earnest continuance of labour, and excellent subtlety of investigation, by which Mr. Aveline and Mr. Clifton Ward have presented you with the marvellous maps and sections of this district, now in course of publication in the Geological Survey. Especially let me, in the strongest terms of grateful admiration, refer to the results which have been obtained by the microscopic observa-

---

* Connected partly, I doubt not, with Ingle, or Inglewood,—brushwood to burn, (hence Justice Inglewood in 'Rob Roy'). I have still omitted 'clough,' or cleugh, given by Johnson in relation to 'dingle,' and constant in Scott, from 'Gander-cleugh' to 'Buc(k)-cleugh.'

tions of minerals instituted by Mr. Sorby, and carried out indefatigably by Mr. Clifton Ward, forming the first sound foundations laid for the solution of the most secret problems of geology.

5. But while I make this most sincere acknowledgment of what has been done by these gentlemen, and by their brother geologists in the higher paths of science, I must yet in all humility lament that this vast fund of gathered knowledge is every bit of it, hitherto, beyond you and me. Dealing only with infinitude of space and remoteness of time, it leaves us as ignorant as ever we were, or perhaps, in fancying ourselves wiser, even more ignorant, of the things that are near us and around,—of the brooks that sing to us, the rocks that guard us, and the fields that feed.

6. To-night, therefore, I am here for no other purpose than to ask the simplest questions; and to win your interest, if it may be, in pleading with our geological teachers for the answers which as yet they disdain to give.

Here, in your long winding dale of the Kent,—and over the hills, in my little nested dale of the Yew,—will you ask the geologist, with me, to tell us how their pleasant depth was opened for us, and their lovely borders built. For, as yet, this is all that we are told concerning them, by accumulated evidence of geology, as collected in this summary at the end of the first part of Mr. Clifton Ward's volume on the geology of the lakes :—

"The most ancient geologic records in the district indicate marine conditions with a probable proximity of land. Submarine volcanoes broke out during the close of this period, followed by an elevation of land, with continued volcanic eruptions of which perhaps the present site of Keswick was one of the chief centres. Depression of the volcanic district then ensued beneath the sea, with the probable cessation of volcanic activity; much denudation was effected; another slight volcanic outburst accompanied the formation of the Coniston Limestone, and then the old deposits of Skiddaw Slate and volcanic material were buried thousands of feet deep beneath strata formed in an upper Silurian sea. Next followed an immensely long period of elevation, accompanied by disturbance and alteration of the rocks, and by a prodigious amount of marine and atmospheric denudation. A subsequent depression, to a considerable extent, marked the coming on of the Carboniferous epoch, heralded however, in all likelihood, by a period of more or less intense cold. Then for succeeding ages, the district elevated high above the surrounding seas of later times, underwent that large amount of sub-aerial denudation which has resulted in the formation of our beautiful English Lake-country."

7. The only sentence in this passage of the smallest service to us, at present, is that stating the large amount of 'sub-aerial denudation' which formed our beautiful country.

Putting the geological language into simple English, that means that your dales and hills were produced by being 'rubbed down in the open air,'—rubbed down, that is to say, in the manner in which people are rubbed down after a Turkish bath, so as to have a good deal of their skin taken off them. But observe, it would be just as rational to say that the beauty of the human form was owing to the immemorial and continual use of the flesh-brush, as that we owe the beauty of our mountains to the mere fact of their having been rubbed away. No quantity of stripping or denuding will give beauty when there is none to denude;—you cannot rub a statue out of a sandbank, or carve the Elgin frieze with rottenstone for a chisel, and chance to drive it.

8. We have to ask then, first, what material there was here to carve; and then what sort of chisels, and in what workman's hand, were used to produce this large piece of precious chasing or embossed work, which we call Cumberland and West*e*-more-laud.

I think we shall get at our subject more clearly, however, by taking a somewhat wider view of it than our own dales permit, and considering what 'sub-aerial denudation' means, on the surface of the world, instead of in Westmoreland only.

9. Broadly, therefore, we have, forming a great part of that surface, vast plains or steppes, like the levels of France, and lowlands of England, and prairies of America, composed mostly of horizontal beds of soft stone or

gravel. Nobody in general talks of these having been rubbed down; so little, indeed, that I really do not myself know what the notions of geologists are on the matter. They tell me that some four-and-twenty thousand feet or so of slate—say, four miles thick of slate—must have been taken off the top of Skiddaw to grind that into what it is; but I don't know in the least how much chalk or freestone they think has been ground off the East Cliff at Brighton, to flatten that into what it is. They tell me that Mont Blanc must have been three times as high as he is now, when God, or the affinity of atoms, first made him; but give me no idea whatever how much higher the shore of the Adriatic was than it is now, before the lagoon of Venice was rubbed out of it.

10. Collecting and inferring as best I can, it seems to me they mean generally that all the mountains were much higher than they are now, and all the plains lower; and that what has been scraped off the one has been heaped on to the other: but that is by no means generally so; and in the degree in which it is so, hitherto has been unexplained, and has even the aspect of being inexplicable.

I don't know what sort of models of the district you have in the Museum, but the kind commonly sold represent the entire mountain surface merely as so much sandheap washed into gutters. It is totally impossible for your youth, while these false impressions are conveyed by the cheap tricks of geographical manufacture, to approach the problems of mountain form under any sense of their real conditions: while even advanced geologists are too much in the habit of thinking that every mountain mass may be considered as a heap of homogeneous clay, which some common plough has fretted into similar clods.

But even to account for the furrows of a field you must ask for plough and ploughman. How much more to account for the furrows of the adamantine rock. Shall one plough *there* with oxen?

I will ask you, therefore, to-night, to approach this question in its first and simplest terms, and to examine the edge of the weapon which is supposed to be still at work. The streamlets of the dale seem yet in many places to be excavating their glens as they dash down them,—or deepening the pools under their cascades. Let us in such simple and daily visible matters consider more carefully what are the facts.

11. Towards the end of July, this last summer, I was sauntering among the fern, beside the bed of the Yewdale stream, and stopped, as one does instinctively, at a place where the stream stopped also,—bending itself round in a quiet brown eddy under the root of an oak tree

How many thousand thousand times have I not stopped to look down into the pools of a mountain stream,—and yet never till that day had it occurred to me to ask how the pools came there. As a matter of course, I had always said to myself, there must be deep places and shallow ones,—and where the water is deep there is an eddy, and where it is shallow there is a ripple,—and what more is there to say about it?

However, that day, having been of late in an interrogative humour about everything, it did suddenly occur to me to ask why the water should be deep there, more than anywhere else. This pool was at a bend of the stream, and rather a wide part of it; and it seemed to me that, for the most part, the deep pools I recollected *had* been at bends of streams, and in rather wide parts of them;—with the accompanying condition of slow circular motion in the water; and also, mostly under steep banks.

12. Gathering my fifty years' experience of brooks, this seemed to me a tenable generalization, that on the whole, where the bank was steepest, and one was most likely to tumble in, one was least likely to get out again.

And that gloomily slow and sullen motion on the surface, as if the bubbles were unwillingly going round in a mill,—this also I recollected as a usual condition of the deeper water,—*so* usual, indeed, that (as I say) I never once before had reflected upon it as the least odd. Whereas now, the thought struck me as I looked, and struck me harder as I looked longer, If the *bubbles* stay at the top, why don't the *stones* stay at the bottom? If, when I throw in a stick here in the back eddy at the surface, it keeps spinning slowly round and round, and never goes down-stream—am I to expect that when I throw a stone into the same eddy, it will be immediately lifted by it out of the hole and carried away? And yet unless the water at the bottom of the hole has this power of lifting stones out of it, why is the hole not filled up?

13. Coming to this point of the question, I looked up the beck, and down. Up the beck, above the pool, there was a shallow rapid over innumerable stones of all sizes: and down the beck, just below the pool, there was a ledge of rock against which the stream had deposited a heap of rolled shingle, and over the edges of which it flowed in glittering tricklets, so shallow that a child of four years old might have safely waded across; and between the loose stones above in the steep rapid, and the ledge of rock below—which seemed put there expressly for them to be lodged against—here was this deep, and wide, and quiet, pool.

So I stared at it, and stared; and the more I stared, the less I understood it. And if you like, any of you may easily go and stare too, for the pool in question is visible enough from the coach-road, from Mr. Sly's Waterhead Inn, up to Tilberthwaite. You turn to the right from the bridge at Mr. Bowness's smithy, and then in a quarter of a mile you may look over the roadside wall into this quiet recess of the stream, and consider of many things. For, observe, if there were anything out of the way in the pool—I should not send you to look at it. I mark it only for one of myriads such in every mountain stream that ever trout leaped or ripple laughed in.

And beside it, as a type of all its brother deeps, these following questions may be wisely put to yourselves.

14. First—How are any of the pools kept clear in a stream that carries shingle? There is some power the water has got of lifting it out of the deeps hitherto unexplained—unthought of. Coming down the rapid in a rage, it drops the stones, and leaves them behind; coming to the deep hole, where it seems to have no motion, it picks them up and carries them away in its pocket. Explain that.

15. But, secondly, beside this pool let us listen to the wide murmuring geological voice, telling us—"To sub-aerial denudation you owe your beautiful lake scenery"! —Then, presumably, Yewdale among the rest?—Therefore we may look upon Yewdale as a dale sub-aerially denuded. That is to say, there was once a time when no dale was there, and the process of denudation has excavated it to the depth you see.

16. But now I can ask, more definitely and clearly, With what chisel has this hollow been hewn for us? Of course, the geologist replies, by the frost, and the rain, and the decomposition of its rocks. Good; but though frost may break up, and the rain wash down, there must have been somebody to cart away the rubbish, or still you would have had no Yewdale. Well, of course, again the geologist answers, the streamlets are the carters; and this stream past Mr. Bowness's smithy is carter-in-chief.

17. How many cartloads, then, may we suppose the

stream has carried past Mr. Bowness's, before it carted away all Yewdale to this extent, and cut out all the northern side of Wetherlam, and all that precipice of Yewdale Crag, and carted all the rubbish first into Coniston Lake, and then out of it again, and so down the Crake into the sea? Oh, the geologists reply, we don't mean that the little Crake did all that. Of course it was a great river full of crocodiles a quarter of a mile long; or it was a glacier five miles thick, going ten miles an hour; or a sea of hot water fifty miles deep,—or,—something of that sort. Well, I have no interest, myself, in *any*thing of that sort: and I want to know, here, at the side of my little puzzler of a pool, whether there's any sub-aerial denudation going on still, and whether this visible Crake, though it can only do little, does *any*thing. Is it carrying stones at all, now, past Mr. Bowness's? Of course, reply the geologists; don't you see the stones all along it, and doesn't it bring down more every flood? Well, yes; the delta of Coniston Waterhead may, perhaps, within the memory of the oldest inhabitant, or within the last hundred years, have advanced a couple of yards or so. At that rate, those two streams, considered as navvies, are proceeding with the works in hand;—to that extent they are indeed filling up the lake, and to that extent subaerially denuding the mountains. But now, I must ask your attention very closely: for I have a strict bit of logic to put before you, which the best I can do will not make clear without some helpful effort on your part.

18. The streams, we say, by little and little, are filling up the lake. They did not cut out the basin of that. Something else must have cut out that, then, before the streams began their work. Could the lake, then, have been cut out all by itself, and none of the valleys that lead to it? Was it punched into the mass of elevated ground like a long grave, before the streams were set to work to cut Yewdale down to it?

19. You don't for a moment imagine that. Well, then, the lake and the dales that descend with it, must have been cut out together. But if the lake not by the streamlets, then the dales not by the streamlets? The streamlets are the consequence of the dales then,—not the causes; and the sub-aerial denudation to which you owe your beautiful lake scenery, must have been something, not only different from what is going on now, but, in one half of it at least, *contrary* to what is going on now. Then, the lakes which are now being filled up, were being cut down; and as probably, the mountains now being cut down, were being cast up.

20. Don't let us go too fast, however. The streamlets are now, we perceive, filling up the big lake. But are they not, then, also filling up the little ones? If they don't cut Coniston water deeper, do you think they are cutting Mr. Marshall's tarns deeper? If not Mr. Marshall's tarns deeper, are they cutting their own little pools deeper? This pool by which we are standing—we have seen it is inconceivable how it is not filled up,—much

more it is inconceivable that it should be cut deeper down. You can't suppose that the same stream which is filling up the Coniston lake below Mr. Bowness's, is cutting out another Coniston lake above Mr. Bowness's? The truth is that, above the bridge as below it, and from their sources to the sea, the streamlets have the same function, and are filling, not deepening, alike lake, tarn, pool, channel, and valley.

21. And that being so, think how you have been misled by seeking knowledge far afield, and for vanity's sake, instead of close at home, and for love's sake. You must go and see Niagara, must you?—and you will brick up and make a foul drain of the sweet streamlet that ran past your doors. And all the knowledge of the waters and the earth that God meant for you, flowed with it, as water of life.

Understand, then, at least, and at last, to-day, Niagara is a vast Exception—and Deception. The true cataracts and falls of the great mountains, as the dear little cascades and leaplets of your own rills, fall where they fell of old; —that is to say, wherever there's a hard bed of rock for them to jump over. They don't cut it away—and they can't. They do form pools *beneath* in a mystic way,—they excavate them to the depth which will break their fall's force—and then they excavate no more.*

We must look, then, for some other chisel than the

* Else every pool would become a well, of continually increasing depth.

streamlet; and therefore, as we have hitherto interrogated the waters at their work, we will now interrogate the hills, in their patience.

22. The principal flank of Yewdale is formed by a steep range of crag, thrown out from the greater mass of Wetherlam, and known as Yewdale Crag.

It is almost entirely composed of basalt, or hard volcanic ash; and is of supreme interest among the southern hills of the lake district, as being practically the first rise of the great mountains of England, out of the lowlands of England.

And it chances that my own study window being just opposite this crag, and not more than a mile from it as the bird flies, I have it always staring me, as it were, in the face, and asking again and again, when I look up from writing any of my books,—"How did *I* come here?"

I wrote that last sentence hurriedly, but leave it—as it was written; for, indeed, however well I know the vanity of it, the question is still sometimes, in spite of my best effort, put to me in that old form by the mocking crags, as by a vast couchant Sphinx, tempting me to vain labour in the inscrutable abyss.

But as I regain my collected thought, the mocking question ceases, and the divine one forms itself, in the voice of vale and streamlet, and in the shadowy lettering of the engraven rock.

"Where wast thou when I laid the foundation of the earth?—declare, if thou hast understanding."

23. How Yewdale Crags came there, I, for one, will no more dream, therefore, of knowing, than the wild grass can know, that shelters in their clefts. I will only to-night ask you to consider one more mystery in the things they have suffered since they came.

You might naturally think, following out the idea of 'sub-aerial denudation,' that the sudden and steep rise of the crag above these softer strata was the natural consequence of its greater hardness; and that in general the district was only the remains of a hard knot or kernel in the substance of the island, from which the softer superincumbent or surrounding material had been more or less rubbed or washed away.*

24. But had that been so, one result of the process must have been certain—that the hard rocks would have resisted more than the soft; and that in some distinct proportion and connection, the hardness of a mountain would be conjecturable from its height, and the whole surface of the district more or less manifestly composed of hard bosses or ridges, with depressions between them in softer materials. Nothing is so common, nothing so clear, as this condition, on a small scale, in every weathered rock.

---

* The most wonderful piece of weathering, in all my own district, is on a *projecting* mass of intensely hard rock on the eastern side of Goat's Water. It was discovered and shown to me by my friend the Rev. F. A. Malleson; and exactly resembles deep ripple-marking, though nothing in the grain of the rock indicates its undulatory structure.

Its quartz, or other hard knots and veins, stand out from the depressed surface in raised walls, like the divisions between the pits of Dante's eighth circle,—and to a certain extent, Mr. Ward tells us, the lava dykes, either by their hardness or by their decomposition, produce walls and trenches in the existing surface of the hills. But these are on so small a scale, that on this map they cannot be discernibly indicated; and the quite amazing fact stands out here in unqualified and indisputable decision, that by whatever force these forms of your mountains were hewn, it cut through the substance of them, as a sword-stroke through flesh, bone, and marrow, and swept away the masses to be removed, with as serene and indiscriminating power as one of the shot from the Devil's great guns at Shoeburyness goes through the oak and the iron of its target.

25. It is with renewed astonishment, whenever I take these sections into my hand, that I observe the phenomenon itself; and that I remember the persistent silence of geological teachers on this matter, through the last forty years of their various discourse. In this shortened section, through Bowfell to Brantwood, you go through the summits of three first-rate mountains down to the lowland moors: you find them built, or heaped; barred, or bedded; here with forged basalt, harder than flint and tougher than iron,—there, with shivering shales that split themselves into flakes as fine as puff-paste, and as brittle as shortbread. And behold, the hewing tool of the Mas-

REESE LIBRARY
OF THE
UNIVERSITY

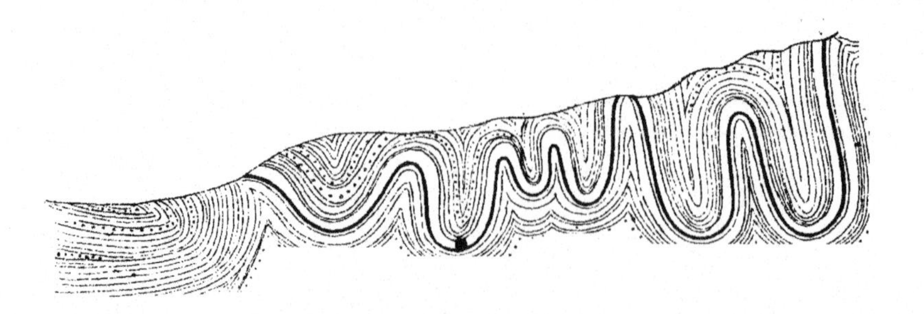

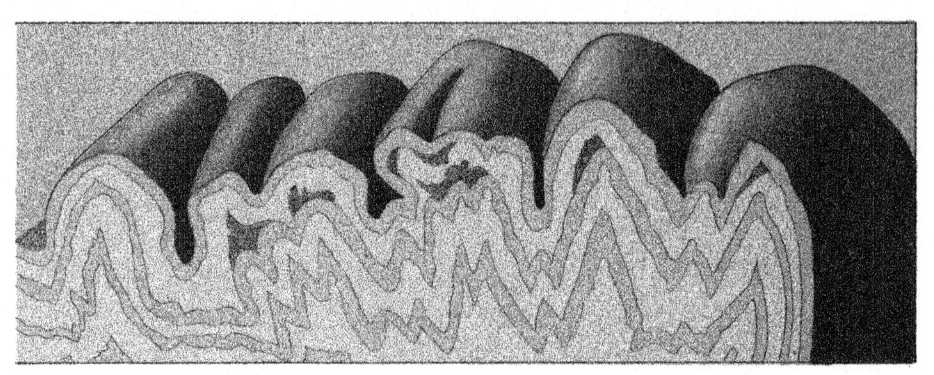

right, simple.

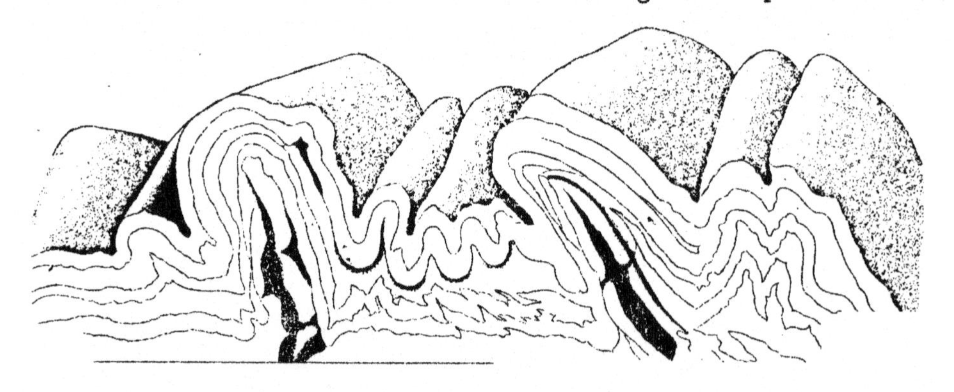

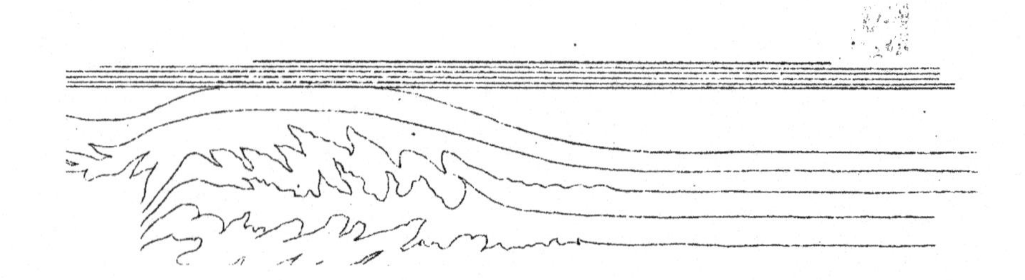

ter Builder sweeps along the forming lines, and shapes the indented masses of them, as a draper's scissors shred a piece of striped sarsnet!

26. Now do but think a little of the wonderfulness in this. If the process of grinding was slow, why don't the hard rocks project? If swift, what kind of force must it have been? and why do the rocks it tore show no signs of rending? Nobody supposes it was indeed swift as a sword or a cannon-ball; but if not, why are the rocks not broken? Can you break an oak plank and leave no splinters, or cut a bed of basalt a thousand feet thick like cream-cheese.

But you suppose the rocks were soft when it was done. Why don't they squeeze, then?

Make Dover cliffs of baker's dough, and put St. Paul's on the top of them,—won't they give way somewhat, think you? and will you then make Causey Pike of clay, and heave Scawfell against the side of it; and yet shall it not so much as show a bruise?

Yet your modern geologists placidly draw the folded beds of the Skiddaw and Causey Pike slate, *first*, without observing whether the folds they draw are *possible* folds in anything; and, *secondly*, without the slightest suggestion of sustained pressure, or bruise, in any part of them

27. I have given in my diagram, (Plate VI., Fig. 1,) the section, attributed, in that last issued by the Geological Survey, to the contorted slates of Maiden Moor, be-

tween Causey Pike and the erupted masses of the central mountains. Now, for aught I know, those contorsions may be truly represented;—but if so, they are not contortions by lateral pressure. For, first, they are impossible forms in any substance whatever, capable of being contorted; and, secondly, they are doubly impossible in any substance capable of being squeezed.

Impossible, I say, first in any substance capable of being contorted. Fold paper, cloth, leather, sheets of iron,—what you will, and still you can't *have the folded bed at the top double the length of that at the bottom.* But here, I have measured the length of the upper bed, as compared with that of the lower, and it is twenty miles, to eight miles and a half.

Secondly, I say, these are impossible folds in any substance capable of being squeezed, for every such substance will change its form as well as its direction under pressure. And to show you how such a substance does actually behave, and contort itself under lateral pressure, I have prepared the sections Figures 2, 3, and 4.

28. I have just said, you have no business to seek knowledge far afield, when you can get it at your doors. But more than that, you have no business to go outside your doors for it, when you can get it in your parlour. And it so happens that the two substances which, while the foolish little king was counting out his money, the wise little queen was eating in the parlour, are precisely the two substances beside which wise little queens, and

kings, and everybody else, may also think, in the parlour,—Bread and honey. For whatever bread, or at least dough, will do under pressure, ductile rocks, in their proportion, must also do under pressure; and in the manner that honey will move, poured upon a slice of them,—in that manner, though in its own measure, ice will move, poured upon a bed of them. Rocks, no more than pie-crust, can be rolled out without squeezing them thinner; and flowing ice can no more excavate a valley, than flowing treacle a teaspoon.

29. I said just now, Will you dash Scawfell against Causey Pike?

I take, therefore, from the Geological Survey the section of the Skiddaw slates, which continue the mass of Causey Pike under the Vale of Newlands, to the point where the volcanic mass of the Scawfell range thrusts itself up against them, and laps over them. They are represented, in the section, as you see, (Plate VI., Fig. 1;) and it has always been calmly assumed by geologists that these contortions were owing to lateral pressure.

But I must beg you to observe that since the uppermost of these beds, if it were straightened out, would be more than twice the length of the lower ones, you could only obtain that elongation by squeezing the upper bed more than the lower, and making it narrower where it is elongated. Now, if this were indeed at the surface of the ground, the geologists might say the upper bed had been thrown up because there was less weight on it. But, by their own accounts, there were five miles thick of rocks on the top of all this when it was bent. So you could not have made one bed tilt up, and another stay down; and the structure is evidently an impossible one.

30. Nay, answer the surveyors, impossible or not, it is there. I partly, in pausing, myself doubt its being there. This looks to me an ideal, as well as an impossible, undulation.

But if it is indeed truly surveyed, then assuredly, whatever it may be owing to, it is not owing to lateral pressure.

That is to say, it may be a crystalline arrangement assumed under pressure, but it is assuredly *not* a form assumed by ductile substance under mechanical force. Order the cook to roll out half a dozen strips of dough, and to stain three of them with cochineal. Put red and white alternately one above the other. Then press them in any manner you like; after pressure, a wetted carving knife will give you quite unquestionable sections, and you see the results of three such experiments in the lower figures of the plate.

31. Figure 2 represents the simplest possible case. Three white and three red dough-strips were taken, a red one uppermost, (for the pleasure of painting it afterwards)! They were left free at the top, enclosed at the sides, and then reduced from a foot to six inches in length, by pressure from the right. The result, you see,

is that the lower bed rises into sharpest gables; the upper ones are rounded softly. But in the geological section it is the upper bed that rises, the lower keeps down! The second case is much more interesting. The pastes were arranged in the same order, but bent up a little, to begin with, in two places, before applying the pressure. The result was, to my own great surprise, that at these points of previous elevation, the lower bed first became quite straight by tension as it rose, and then broke into transverse faults.

32. The third case is the most interesting of all. In this case, a roof of slate was put over the upper bed, allowing it to rise to some extent only, and the pressure was applied to the two lower beds only.* The upper bed of course exuded backwards, giving these flame-like forms, of which afterwards I got quite lovely complications by repeated pressures. These I must reserve for future illustration, concluding to-night, if you will permit me, with a few words of general advice to the younger members of this society, formed as it has been to trace for itself a straight path through the fields of literature, and over the rocks of science.

33. First.—Whenever you write or read English, write it pure, and make it pure if ill written, by avoiding all unnecessary foreign, especially Greek, forms of words

* Here I had to give the left-hand section, as it came more neatly. The wrinkled mass on the left coloured brown represents the pushing piece of wood, at the height to which it was applied.

yourself, and translating them when used by others. Above all, make this a practice in science. Great part of the supposed scientific knowledge of the day is simply bad English, and vanishes the moment you translate it.

There is a farther very practical reason for avoiding all vulgar Greek-English. Greece is now a kingdom, and will I hope remain one, and its language is now living. The ship-chandler, within six doors of me on the quay at Venice, had indeed a small English sign—calling himself Ship-Chandler; but he had a large and practically more serviceable, Greek one, calling himself a "*προμηθεττης τῶν πλοιων*." Now when the Greeks want a little of your science, as in very few years they must, if this absurd practice of using foreign languages for the clarification of scientific principle still holds, what you, in compliment to Greece, call a 'Dinotherium,' Greece, in compliment to you, must call a 'Nasty-beastium,'—and you know that interchange of compliments can't last long.

34. II. Observe generally that all knowledge, little or much, is dangerous, in which your progress is likely to be broken short by any strict limit set to the powers of mortals: while it is precisely that kind of knowledge which provokes vulgar curiosity, because it seems so far away; and idle ambition, because it allows any quantity of speculation, without proof. And the fact is that the greater quantity of the knowledge which modern science

is so saucy about, is only an asses' bridge, which the asses all stop at the top of, and which, moreover, they can't help stopping at the top of; for they have from the beginning taken the wrong road, and so come to a broken bridge—a Ponte Rotto over the river of Death, by which the Pontifex Maximus allows them to pass no step farther.

35. For instance,—having invented telescopes and photography, you are all stuck up on your hobby-horses, because you know how big the moon is, and can get pictures of the volcanoes in it!

But you never can get any more than *pictures* of these, while in your own planet there are a thousand volcanoes which you may jump into, if you have a mind to; and may one day perhaps be blown sky high by, whether you have a mind or not. The last time the great volcano in Java was in eruption, it threw out a stream of hot water as big as Lancaster Bay, and boiled twelve thousand people. That's what I call a volcano to be interested about, if you want sensational science.

36. But if not, and you can be content in the wonder and the power of Nature, without her terror,—here is a little bit of a volcano, close at your very doors—Yewdale Crag, which I think will be quiet for our time,—and on which the anagallis tenella, and the golden potentilla, and the sundew, grow together among the dewy moss in peace. And on the cellular surface of one of the blocks of it, you may find more beauty, and learn

more precious things, than with telescope or photograph from all the moons in the milky way, though every drop of it were another solar system.

I have a few more very serious words to say to the fathers, and mothers, and masters, who have honoured me with their presence this evening, with respect to the influence of these far-reaching sciences on the temper of children.

37. Those parents who love their children most tenderly, cannot but sometimes dwell on the old Christian fancy, that they have guardian angels. I call it an old fancy, in deference to your modern enlightenment in religion; but I assure you nevertheless, in spite of all that illumination, there remains yet some dark possibility that the old fancy may be true: and that, although the modern apothecary cannot exhibit to you either an angel, or an imp, in a bottle, the spiritual powers of heaven and hell are no less now, than heretofore, contending for the souls of your children; and contending with *you*—for the privilege of their tutorship.

38. Forgive me if I use, for the few minutes I have yet to speak to you, the ancient language,—metaphorical, if you will, of Luther and Fenelon, of Dante and Milton, of Goethe and Shakspeare, of St. John and St. Paul, rather than your modern metaphysical or scientific slang: and if I tell you, what in the issue of it you will find is either life-giving, or deadly, fact,—that the fiends and the angels contend with you daily for the spirits of your

children: the devil using to you his old, his hitherto immortal, bribes, of lust and pride; and the angels pleading with you, still, that they may be allowed to lead your babes in the divine life of the pure and the lowly. To enrage their lusts, and chiefly the vilest lust of money, the devils would drag them to the classes that teach them how to get on in the world; and for the better pluming of their pride, provoke their zeal in the sciences which will assure them of there being no God in nature but the gas of their own graves.

And of these powers you may discern the one from the other by a vivid, instant, practical test. The devils always will exhibit to you what is loathsome, ugly, and, above all, dead; and the angels, what is pure, beautiful, and, above all, living.

39. Take an actual, literal instance. Of all known quadrupeds, the unhappiest and vilest, yet alive, is the sloth, having this farther strange devilry in him, that what activity he is capable of, is in storm, and in the night. Well, the devil takes up this creature, and makes a monster of it,—gives it legs as big as hogsheads, claws stretched like the roots of a tree, shoulders like a hump of crag, and a skull as thick as a paving-stone. From this nightmare monster he takes what poor faculty of motion the creature, though wretched, has in its minuter size; and shows you, instead of the clinging climber that scratched and scrambled from branch to branch among the rattling trees as they bowed in storm, only a vast

heap of stony bones and staggering clay, that drags its meat down to its mouth out of the forest ruin. This creature the fiends delight to exhibit to you, but are permitted by the nobler powers only to exhibit to you in its death.*

* The Mylodon. An old sketch, (I think, one of Leech's) in Punch, of Paterfamilias improving Master Tom's mind among the models on the mud-bank of the lowest pond at Sydenham, went to the root of the matter. For the effect, on Master Tom's mind of the living squirrel, compare the following account of the most approved modes of squirrel-hunting, by a clerical patron of the sport, extracted for me by a correspondent, from 'Rabbits: how to rear and manage them; with Chapters on Hares, Squirrels, etc.' S. O. Beeton, 248, Strand, W. C.

"It may be easily imagined that a creature whose playground is the top twigs of tall trees, where no human climber dare venture, is by no means easy to capture—especially as its hearing is keen, and its vision remarkably acute. Still, among boys living in the vicinity of large woods and copses, squirrel-hunting is a favourite diversion, and none the less so because it is seldom attended by success. 'The only plan,' says the Rev. Mr. Wood, 'is to watch the animal until it has ascended an isolated tree, or, by a well-directed shower of missiles, to drive it into such a place of refuge, and then to form a ring round the tree so as to intercept the squirrel, should it try to escape by leaping to the ground and running to another tree. The best climber is then sent in chase of the squirrel, and endeavours, by violently shaking the branches, to force the little animal to loose its hold and fall to the earth. But it is by no means an easy matter to shake a squirrel from a branch, especially as the little creature takes refuge on the topmost and most slender boughs, which even bend under the weight of its own small body, and can in no way be trusted with the weight of a human being. By dint, however, of perseverance, the

40. On the other hand, as of all quadrupeds there is none so ugly or so miserable as the sloth, so, take him for all in all, there is none so beautiful, so happy, so wonderful as the squirrel. Innocent in all his ways, harmless in his food, playful as a kitten, but without cruelty, and surpassing the fantastic dexterity of the monkey, with the grace and the brightness of a bird; the little dark-eyed miracle of the forest glances from branch to branch more like a sunbeam than a living creature: it leaps, and darts, and twines, where it will;—a chamois is slow to it; and a panther, clumsy: grotesque as a gnome, gentle as a fairy, delicate as the silken plumes of the rush, beautiful and strong like the spiral of a fern,—it haunts you, listens for you, hides from you, looks for you, loves you, as if the angel that walks with your children had made it himself for their heavenly plaything.

And this is what *you* do, to thwart alike your child's angel, and his God,—you take him out of the woods into the town,—you send him from modest labour to competitive schooling,—you force him out of the fresh air

---

squirrel is at last dislodged, and comes to the ground as lightly as a snow-flake. Hats, caps, sticks, and all available missiles are immediately flung at the luckless animal as soon as it touches the ground, and it is very probably struck and overwhelmed by a cap. The successful hurler flings himself upon the cap, and tries to seize the squirrel as it lies under his property. All his companions gather round him, and great is the disappointment to find the cap empty, and to see the squirrel triumphantly scampering up some tree where it would be useless to follow it.'"

into the dusty bone-house,—you show him the skeleton of the dead monster, and make him pore over its rotten cells and wire-stitched joints, and vile extinct capacities of destruction,—and when he is choked and sickened with useless horror and putrid air, you let him—regretting the waste of time—go out for once to play again by the woodside; and the first squirrel he sees, he throws a stone at!

Carry, then, I beseech you, this assured truth away with you to-night. All true science begins in the love, not the dissection, of your fellow-creatures; and it ends in the love, not the analysis, of God. Your alphabet of science is in the nearest knowledge, as your alphabet of science is in the nearest duty. "Behold, it is nigh thee, even at the doors." The Spirit of God is around you in the air that you breathe,—His glory in the light that you see; and in the fruitfulness of the earth, and the joy of its creatures, He has written for you, day by day, His revelation, as He has granted you, day by day, your daily bread.

# CHAPTER XIII.

## OF STELLAR SILICA.

1. THE issue of this number of Deucalion has been so long delayed, first by other work, and recently by my illness, that I think it best at once to begin Mr. Ward's notes on Plate V.: reserving their close, with full explanation of their importance and bearing, to the next following number.

GRETA BANK COTTAGE, KESWICK,
*June* 13, 1876.

My dear Sir,—I send you a few notes on the microscopic structure of the three specimens I have had cut. In them I have stated merely what I have seen. There has been much which I did not expect, and still more is there that I don't understand.

I am particularly sorry I have not the time to send a whole series of coloured drawings illustrating the various points; but this summer weather claims my time on the mountain-side, and I must give up microscopic work until winter comes round again.

The minute spherulitic structure—especially along the fine brown lines—was quite a surprise, and I shall hope on some future occasion to see more of this subject. Believe me, yours very truly,

J. Clifton Ward.

P.S.—There seems to be a great difference between the microscopic structure of the specimens now examined and that of the filled-up vesicles in many of my old lavas here, so far as my *limited* examination has gone.

SPECIMEN A.

*No.* 1 *commences at the end of the section farthest from A in specimen.*

1. Transparent zone with irregular curious cavities (not liquid), and a few mossy-looking round spots (brownish).

*Polarization.* Indicating an indefinite semi-crystalline structure. (See note at page 211.)

2. Zone with minute seed-like bodies of various sizes (narrow brownish bands in the specimen of darker and lighter tints).

*a.* Many cavities, and of an indefinite oval form in general.

*b.* The large spherulites (2) are very beautiful, the outer zone (radiate) of a delicate greenish-yellow, the nucleus of a brownish-yellow, and the intermediate zone generally clear.

*c.* A layer of densely packed bodies, oblong, or oval in form.

*d.* Spherulites generally similar to *b*, but smaller,

much more stained of a brownish-yellow, and with more defined nuclei.

*Polarization.* The spherulites show a clearly radiate polarization, with rotation of a dark cross on turning either of the prisms; the intermediate ground shows the irregular semi-crystalline structure.

3. Clear zone, with little yellowish, dark, squarish specks.

*Polarization.* Irregular, semi-crystalline.

4. Row of closely touching spherulites with large nucleus and defined margin, rather furry in character (3). Margins and nuclei brown; intermediate space brownish-yellow.

*Polarization.* Radiate, as in the spherulites 2 *b*.

(This is a short brown band which does not extend down through the whole thickness of the specimen.)

5. Generally clear ground, with a brownish cloudy appearance in parts.

*Polarization.* Indefinite semi-crystalline.

6 *a*. On a hazy ground may be seen the cloudy margins of separately crystalline spaces.

*Polarization.* Definite semi-crystalline.*

---

* By '*indefinite* semi-crystalline' is meant the breaking up of the ground under crossed prisms with sheaves (5) of various colours not clearly margined.

By '*definite* semi-crystalline' is meant the breaking up of the ground under crossed prisms with a mosaic (4) of various colours clearly margined.

By 'semi-crystalline' is meant the interference of crystalline spaces

6 *b.* A clear band with very indefinite polarization.

7. A clearish zone with somewhat of a brown mottled appearance (light clouds of brown colouring matter).

*Polarization.* Indefinite semi-crystalline.

8. Zone of brownish bodies (this is a fine brown line, about the middle of the section in the specimen).

*a.* Yellowish-brown nucleated disks.

*b.* Smaller, scattered, and *generally* non-nucleated disks.

*c.* Generally non-nucleated.

*Polarization.* The disks are too minute to show separate polarization effects, but the ground exhibits the indefinite semi-crystalline.

9. Ground showing indefinite semi-crystalline polarization.

10. Irregular line of furry-looking yellowish disks.

11. Zone traversed by a series of generally parallel and faint lines of a brownish-yellow. These are apparently lines produced by colouring matter alone,—at any rate, not by *visible* disks of any kind.

*Polarization.* Tolerably definite, and limited by the cross lines (6).

12. Dark-brown flocculent-looking matter, as if growing out from a well-defined line, looking like a moss-growth.

---

with one another, so as to prevent a perfect crystalline form being assumed.

13. Defined crystalline interlocked spaces.

*Polarization.* Definite semi-crystalline.

14. Generally, not clearly defined spaces; central part rather a granular look (spaces very small).

*Polarization.* Under crossed prisms breaking up into tolerably definite semi-crystalline spaces.

SPECIMEN B.

B 1. In the slice taken from this side there seems to be frequently a great tendency to spherulitic arrangement, as shown by the polarization phenomena. In parts of the white quartz where the polarization appearance is like that of a mosaic pavement, there is even a semi-spherulitic structure. In other parts there are many spherulites on white and yellowish ground.

Between the many parallel lines of a yellowish colour the polarization (7) effect is that of fibrous coloured sheaves.

Here (8) there is a central clear band (*b*); between it and (*a*) a fine granular line with some larger granules (or very minute spherulites). The part (*a*) is carious, apparently with glass cavities. On the other side of the clear band, at *c*, are half-formed and adherent spherulites; the central (shaded) parts are yellow, and the outer coat, the intermediate portion clearish.

B 2. The slice from this end of the specimen shows the same general structure.

The general tendency to spherulitic arrangement is

well seen in polarized light, dark crosses frequently traversing the curved structures.

Here (in Fig. 9) the portion represented on the left was situated close to the other portion, where the point of the arrow terminates, both crosses appearing together, and revolving in rotation of one of the prisms.

SPECIMEN C.

The slice from this specimen presents far less variety than in the other cases. There are two sets of structural lines—those radiate (10), and those curved and circumferential (11). The latter structure is exceedingly fine and delicate, and not readily seen, even with a high power, owing to the fine radii not being marked out by any colour, the whole section being very clear and white.

A more decidedly, nucleated structure is seen in part 12.

In (13) is a very curious example of a somewhat more glassy portion protruding in finger-like masses into a radiate, clear, and largely spherical portion.

2. These notes of Mr. Clifton Ward's contain the first accurate statements yet laid before mineralogists respecting the stellar crystallization of silica, although that mode of its formation lies at the very root of the structure of the greater mass of amygdaloidal rocks, and of all the most beautiful phenomena of agates. And indeed I have no words to express the wonder with which

I see work like that done by Cloizeaux in the measurement of quartz angles, conclude only in the construction of the marvellous diagram, as subtle in execution as amazing in its accumulated facts,* without the least reference to the conditions of varying energy which produce the spherical masses of chalcedony! He does not even use the classic name of the mineral, but coins the useless one, Geyserite, for the absolutely local condition of the Iceland sinter.

3. And although, in that formation, he went so near the edge of Mr. Clifton Ward's discovery as to announce that "leur masse se compose ellemême de sphères enchâssées dans une sorte de pâte gélatineuse," he not only fails, on this suggestion, to examine chalcedonic structure generally, but arrested himself finally in the pursuit of his inquiry by quietly asserting, "ce genre de structure n'a jamais été rencontré jusqu'ici sur aucune autre variété de silice naturelle ou artificielle,"—the fact being that there is no chalcedonic mass whatever, which does *not* consist of spherical concretions more or less perfect, enclosed in a "pâte gélatineuse."

4. In Professor Miller's manual, which was the basis of Cloizeaux's, chalcedony is stated to appear to be a mixture of amorphous with crystalline silica! and its form taken no account of. Malachite might just as well have been described as a mixture of amorphous with crystalline carbonate of copper!

---

* Facing page 8 of the 'Manuel de Mineralogie.'

5. I will not, however, attempt to proceed farther in this difficult subject until Mr. Clifton Ward has time to continue his own observations. Perhaps I may persuade him to let me have a connected series of figured examples, from pure stellar quartz down to entirely fluent chalcedony, to begin the next volume of Deucalion with;—but I must endeavour, in closing the present one, to give some available summary of its contents, and clearer idea of its purpose; and will only trespass so far on my friend's province as to lay before him, together with my readers, some points noted lately on another kind of semi-crystallization, which bear not merely on the domes of delicate chalcedony, and pyramids of microscopic quartz, but on the far-seen chalcedony of the Dôme du Gonte, and the prismatic towers of the Cervin and dark peak of Aar.

# CHAPTER XIV.

## SCHISMA MONTIUM.

1. THE index closing this column of Deucalion, drawn up by myself, is made as short as possible, and classifies the contents of the volume so as to enable the reader to collect all notices of importance relating to any one subject, and to collate them with those in my former writings. That they need such assemblage from their desultory occurrence in the previous pages, is matter of sincere regret to me, but inevitable, since the writing of a systematic treatise was incompatible with the more serious work I had in hand, on greater subjects. The 'Laws of Fésole' alone might well occupy all the hours I can now permit myself in severe thought. But any student of intelligence may perceive that one inherent cause of the divided character of this book, is its function of advance in parallel columns over a wide field; seeing that, on no fewer than four subjects, respecting which geological theories and assertions have long been alike fantastic and daring, it has shown at least the necessity for revisal of evidence, and, in two cases, for reversal of judgment.

2. I say "it has shown," fearlessly; for at my time of life, every man of ordinary sense, and probity, knows

what he has done securely, and what perishably. And during the last twenty years, none of my words have been set down untried; nor has any opponent succeeded in overthrowing a single sentence of them.

3. But respecting the four subjects above alluded to, (denudation, cleavage, crystallization, and elevation, as causes of mountain form,) proofs of the uncertainty, or even falseness, of current conceptions have been scattered at intervals through my writings, early and late, from 'Modern Painters' to the 'Ethics of the Dust:' and, with gradually increasing wonder at the fury of so-called 'scientific' speculation, I have insisted, year by year, on the undealt with, and usually undreamt of, difficulties which lay at the threshold of secure knowledge in such matters;—trusting always that some ingenuous young reader would take up the work I had no proper time for, and follow out the investigations of which the necessity had been indicated. But I waited in vain; and the rough experiments made at last by myself, a year ago, of which the results are represented in Plate VI. of this volume, are actually the first of which there is record in the annals of geology, made to ascertain the primary physical conditions regulating the forms of contorted strata. The leisure granted me, unhappily, by the illness which has closed my relations with the University of Oxford, has permitted the pursuit of these experiments a little farther; but I must defer account of their results to the following volume, contenting my-

self with indicating, for conclusion of the present one, to what points of doubt in existing theories they have been chiefly directed.

4. From the examination of all mountain ground hitherto well gone over, one general conclusion has been derived, that wherever there are high mountains, there are hard rocks. Earth, at its strongest, has difficulty in sustaining itself above the clouds; and could not hold itself in any noble height, if knitted infirmly.

5. And it has farther followed, in evidence, that on the flanks of these harder rocks, there are yielding beds, which appear to have been, in some places, compressed by them into wrinkles and undulations;—in others, shattered, and thrown up or down to different levels. My own interest was excited, very early in life,* by the forms and fractures in the mountain groups of Savoy; and it happens that the undulatory action of the limestone beds on each shore of the Lake of Annecy, and the final rupture of their outmost wave into the preci-

* I well yet remember my father's rushing up to the drawing-room at Herne Hill, with wet and flashing eyes, with the proof in his hand of the first sentences of his son's writing ever set in type,—'Enquiries on the Causes of the Colour of the Water of the Rhone,' (Magazine of Natural History, September, 1834; (followed next month by 'Facts and Considerations on the Strata of Mont Blanc, and on some Instances of Twisted Strata observable in Switzerland.' I was then fifteen.) My mother and I eagerly questioning the cause of his excitement,—"It's—it's—only *print,*" said he! Alas! how much the 'only' meant!

pice of the Salève, present examples so clear, and so imposing, of each condition of form, that I have been led, without therefore laying claim to any special sagacity, at least into clearer power of putting essential questions respecting such phenomena than geologists of far wider experience, who have confused or amused themselves by collecting facts indiscriminately over vast spaces of ground. I am well convinced that the reader will find more profit in following my restricted steps; and satisfying, or dissatisfying himself, with precision, respecting forms of mountains which he can repeatedly and exhaustively examine.

6. In the uppermost figure in Plate VII., I have enlarged and coloured the general section given rudely above in Figure 1, page 11, of the Jura and Alps, with the intervening plain. The central figure is the southern, and the lowermost figure, which should be conceived as joining it on the right hand, the northern, series of the rocks composing our own Lake district, drawn for me with extreme care by the late Professor Phillips, of Oxford.

I compare, and oppose, these two sections, for the sake of fixing in the reader's mind one essential point of difference among many resemblances; but that they may not, in this comparison, induce any false impressions, the system of colour which I adopt in this plate, and henceforward shall observe, must be accurately understood.

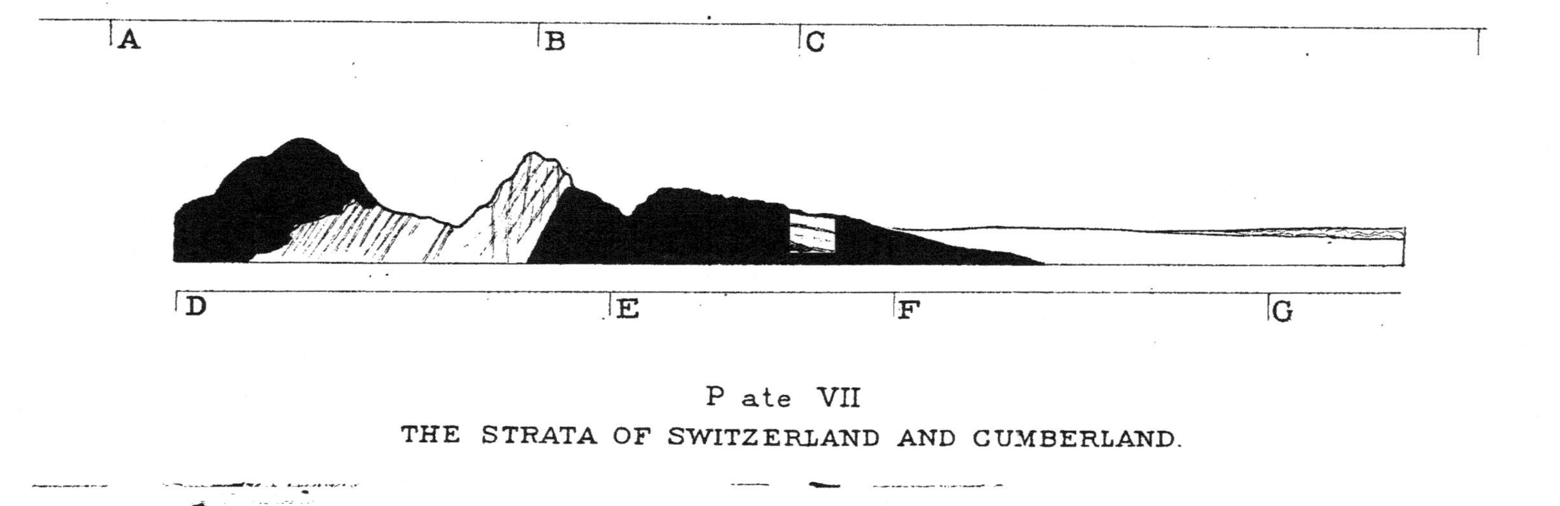


P ate VII

THE STRATA OF SWITZERLAND AND CUMBERLAND.

7. At page 130 above, I gave my reasons for making no endeavour, at the Sheffield Museum, to certify the ages of rocks. For the same reason, in practical sections I concern myself only with their nature and position; and colour granite pink, slate purple, and sandstone red, without inquiring whether the granite is ancient or modern,—the sand trias or pliocene, and the slate Wenlock or Caradoc; but with this much only of necessary concession to recognized method, as to colour with the same tint all rocks which unquestionably belong to the same great geological formation, and vary their mineralogical characters within narrow limits. Thus, since, in characteristic English sections, chalk may most conveniently be expressed by leaving it white, and some of the upper beds of the Alps unquestionably are of the same period, I leave them white also, though their general colour may be brown or grey, so long as they retain cretaceous or marly consistence; but if they become metamorphic, and change into clay slate or gneiss, I colour them purple, whatever their historical relations may be.

8. And in all geological maps and sections given in 'Deucalion,' I shall limit myself to the definition of the twelve following formations by the twelve following colours. It is enough for any young student at first to learn the relations of these great orders of rock and earth:—once master of these, in any locality, he may split his beds into any complexity of finely laminated

chronology he likes;—and if I have occasion to split them for him myself, I can easily express their minor differences by methods of engraving. But, primarily, let him be content in the recognition of these twelve territories of Demeter, by this following colour heraldry:—

9.

| | |
|---|---|
| 1. Granite will bear in the field, | ———— Rose-red. |
| 2. Gneiss and mica-slate | ———— Rose-purple. |
| 3. Clay-slate | ———— Violet-purple. |
| 4. Mountain limestone | ———— Blue. |
| 5. Coal measures and millstone grit | ———— Grey. |
| Jura limestone | ———— Yellow. |
| Chalk | ———— White. |
| Tertiaries forming hard rock | ———— Scarlet. |
| 6. Tertiary sands and clays | ———— Tawny |
| 10. Eruptive rocks not definitely volcanic | ———— Green. |
| 11. Eruptive rocks, definitely volcanic, but at rest | ———— Green, spotted red. |
| 12. Volcanic rocks, active | ———— Black, spotted red. |

10. It will at once be seen by readers of some geological experience, that approximately, and to the degree possible, these colours are really characteristic of the several formations; and they may be rendered more so by a little care in modifying the tints. Thus the 'scarlet' used for the tertiaries may be subdued as much as we please, to what will be as near a sober brown as we can venture without confusing it with the darker

shades of yellow; and it may be used more pure to represent definitely red sandstones or conglomerates· while, again, the old red sands of the coal measures may be extricated from the general grey by a tint of vermilion which will associate them, as mineral substances, with more recent sand. Thus in the midmost section of Plate VII. this colour is used for the old red conglomerates of Kirby Lonsdale. And again, keeping pure light blue for the dated mountain limestones, which are indeed, in their emergence from the crisp turf of their pastures, grey, or even blue in shade, to the eye, a deeper blue may be kept for the dateless limestones which are associated with the metamorphic beds of the Alps; as for my own Coniston Silurian limestone, which may be nearly as old as Skiddaw.

11. The colour called 'tawny,' I mean to be as nearly that of ripe wheat as may be, indicating arable land, or hot prairie; while, in maps of northern countries, touched with points of green, it may pass for moorland and pasture: or, kept in the hue of pale vermilion, it may equally well represent desert alluvial sand. Finally, the avoidance of the large masses of fierce and frightful scarlet which render modern geological maps intolerable to a painter's sight, (besides involving such geographical incongruities as the showing Iceland in the colour of a red-hot coal;) and the substitution over all volcanic districts, of the colour of real greenstone, or serpentine, for one which resembles neither these, nor the general tones

of dark colour either in lava or cinders, will certainly render all geological study less injurious to the eyesight, and less harmful to the taste.

12. Of the two sections in Plate VII., the upper one is arranged from Studer, so as to exhibit in one view the principal phenomena of Alpine structure according to that geologist. The cleavages in the central granite mass are given, however, on my own responsibility, not his. The lower section was, as aforesaid, drawn for me by my kind old friend Professor Phillips, and is, I doubt not, entirely authoritative. In all great respects, the sections given by Studer are no less so; but they are much ruder in drawing, and can be received only as imperfect summaries—perhaps, in their abstraction, occasionally involving some misrepresentation of the complex facts. For my present purposes, however, they give me all the data required.

13. It will instantly be seen, on comparing the two groups of rocks, that although nearly similar in succession, and both suggesting the eruptive and elevatory force of the granitic central masses, there is a wide difference in the manner of the action of these on the strata lifted by them. In the Swiss section, the softer rocks seem to have been crushed aside, like the ripples of water round any submersed object rising to the surface. In the English section, they seem to have undergone no such torsion, but to be lifted straight, as they lay, like the timbers of a gabled roof. It is true that, on the larger

scale of the Geological Survey, contortions are shown at most of the faults in the Skiddaw slate; but, for the reasons already stated, I believe these contortions to be more or less conventionally represented; and until I have myself examined them, will not modify Professor Phillips' drawing by their introduction.

Some acknowledgment of such a structure is indeed given by him observably in the dark slates on the left in the lowermost section; but he has written under these undulatory lines "quartz veins," and certainly means them, so far as they are structural, to stand only for ordinary gneissitic contortion in the laminated mass, and not for undulating strata.

14. Farther. No authority is given me by Studer for dividing the undulatory masses of the outer Alps by any kind of cleavage-lines. Nor do I myself know examples of fissile structure in any of these mountain masses, unless where they are affected by distinctly metamorphic action, in the neighbourhood of the central gneiss or mica-schist. On the contrary, the entire courses of the Cumberland rock, from Kirby Lonsdale to Carlisle, are represented by Professor Phillips as traversed by a perfectly definite and consistent cleavage throughout, dipping steeply south, in accurately straight parallel lines, and modified only, in the eruptive masses, by a vertical cleavage, characterizing the pure granite centres.

15. I wish the reader to note this with especial care, because the cleavage of secondary rock has been lately

attributed, with more appearance of reason than modern scientific theories usually possess, to lateral pressure, acting in a direction perpendicular to the lamination. It seems, however, little calculated to strengthen our confidence in such an explanation, to find the Swiss rocks, which appear to have been subjected to a force capable of doubling up leagues of them backwards and forwards like a folded map, wholly without any resultant schistose structure; and the English rocks, which seem only to have been lifted as a raft is raised on a wave, split across, for fifty miles in succession, by foliate structures of the most perfect smoothness and precision.

16. It might indeed be alleged, in deprecation of this objection, that the dough or batter of which the Alps were composed, mostly calcareous, did not lend itself kindly to lamination, while the mud and volcanic ashes of Cumberland were of a slippery and unctuous character, easily susceptible of rearrangement under pressure. And this view receives strong support from the dextrous experiment performed by Professor Tyndall in 1856, and recorded, as conclusive, in 1872,* wherein, first warming some wax, then pressing it between two pieces of glass, and finally freezing it, he finds the congealed mass delicately laminated; and attributes its lamination to the "lateral sliding of the particles over each other." * But with his usual, and quite unrivalled, incapacity of follow-

* 'Forms of Water,' King and Co., 1872, p. 190.

ing out any subject on the two sides of it, he never tells us, and never seems to have asked himself, how *far* the wax was flattened, and how far, therefore, its particles had been forced to slide;—nor, during the sixteen years between his first and final record of the experiment, does he seem ever to have used any means of ascertaining whether, under the observed conditions, real compression of the substance of the wax had taken place at all! For if not, and the form of the mass was only altered from a lump to a plate, without any increase of its density, a less period for reflection than sixteen years might surely have suggested to Professor Tyndall the necessity, in applying his result to geological matters, of providing mountains which were to be squeezed in one direction, with room for expansion in another.

17. For once, however, Professor Tyndall is not without fellowship in his hesitation to follow the full circumference of this question. Among the thousands of passages I have read in the works even of the most careful and logical geologists,—even such as Humboldt and De Saussure,—I remember *not one* distinct statement * of

---

* As these sheets are passing through the press, I received the following most important note from Mr. Clifton Ward: "With regard to the question whether cleavage is necessarily followed by a reduction in bulk of the body cleaved, the following cases may help us to form an opinion. *Crystalline* volcanic rocks (commonly called trap), as a rule, are not cleaved, though the beds, uncrystalline in character, above and below them, may be. When, however, a trap is highly

the degree in which they supposed the lamination of any given rock to imply real increase of its density, or only the lateral extension of its mass.

18. And the student must observe that in many cases lateral extension of mass is precisely avoided by the very positions of rocks which are supposed to indicate the pressure sustained. In Mr. Woodward's experiment with sheets of paper, for instance, (above quoted, p. 17,*) there is neither increase of density nor extension of mass, in the sheets of paper. They remain just as thick as they were,—just as long and broad as they were. They are only altered in direction, and no more compressed, as they bend, than a flag is compressed by the wind that

---

vesicular, it is sometimes well cleaved. May we not, therefore, suppose that in a rock, *wholly* crystalline, the particles are too much interlocked to take up new positions? In a purely fragmentary rock, however, the particles seem to have more freedom of motion; their motion under pressure leads to a new and more parallel arrangement of particles, each being slightly flattened or pulled out along the planes of new arrangement. This, then, points to a diminution of bulk at any rate in a direction at right angles to the planes of cleavage. The tendency to new arrangement of particles *under pressure points to accommodation under altered circumstances of space.* In rocks composed of fragments, the interspaces, being for the most part larger than the intercrystalline spaces of a trap rock, more freely allow of movement and new arrangement."

* There is a double mistake in the fourth line from the top in that page. I meant to have written, "from a length of four inches into the length of one inch,"—but I believe the real dimensions should have been "a foot crushed into three inches."

waves it. In my own experiments with dough, of course the dough was no more compressible than so much water would have been. Yet the language of the geologists who attribute cleavage to pressure might usually leave their readers in the notion that clay can be reduced like steam; and that we could squeeze the sea down to half its depth by first mixing mud with it! Else, if they really comprehended the changes of form rendered necessary by proved directions of pressure, and did indeed mean that the paste of primitive slate had been 'flattened out' (in Professor Tyndall's words) as a cook flattens out her pastry-crust with a rolling-pin, they would surely sometimes have asked themselves,—and occasionally taken the pains to tell their scholars,—where the rocks in question had been flattened to. Yet in the entire series of Swiss sections (upwards of a hundred) given by Studer in his Alpine Geology, there is no hint of such a difficulty having occurred to him;—none, of his having observed any actual balance between diminution of bulk and alteration of form in contorted beds;—and none, showing any attempt to distinguish mechanical from crystalline foliation. The cleavages are given rarely in any section, and always imperfectly.

19. In the more limited, but steadier and closer, work of Professor Phillips on the geology of Yorkshire, the solitary notice of "that very obscure subject, the cleavage of slate" is contained in three pages, (5 to 8 of the first chapter,) describing the structure of a single quarry,

in which the author does not know, and cannot eventually discover, whether the rock is stratified or not! I respect, and admire, the frankness of the confession; but it is evident that before any affirmation of value, respecting cleavages, can be made by good geologists, they must both ascertain many laws of pressure in viscous substances at present unknown; and describe a great many quarries with no less attention than was given by Professor Phillips to this single one.

20. The experiment in wax, however, above referred to as ingeniously performed by Professor Tyndall, is not adduced in the "Forms of water" for elucidation of cleavage in rocks, but of riband structure in ice—(of which more presently). His first display of it, however, was I believe in the lecture delivered in 1856 at the Royal Institution,—this, and the other similar experiments recorded in the Appendix to the 'Glaciers of the Alps,' being then directed mainly to the confusion of Professor Sedgwick, in that the Cambridge geologist had —with caution—expressed an opinion that cleavage was a result of crystallization under polar forces.

21. Of that suggestion Professor Tyndall complimentarily observed that "it was a bold stretch of analogies," and condescendingly—that "it had its value,—it has drawn attention to the subject." Presently, translating this too vulgarly intelligible statement into his own sublime language, he declares of the theory in debate that it, and the like of it, are "a dynamic power which oper-

ates against intellectual stagnation." How a dynamic power differs from an undynamic one,—and, presumably, also, a potestatic dynamis from an unpotestatic one;) and how much more scientific it is to say, instead of—that our spoon stirs our porridge,—that it "operates against the stagnation" of our porridge, Professor Tyndall trusts the reader to recognize with admiration. But if any stirring or skimming, or other operation of a duly dynamic character, could have clarified from its scum of vanity the pease-porridge of his own wits, Professor Tyndall would have felt that men like the Cambridge veteran,—one of the very few modern men of science who possessed real genius,—stretch no analogies farther than they will hold; and, in this particular case, there were two facts, familiar to Sedgwick, and with which Professor Tyndall manifests no acquaintance, materially affecting every question relating to cleavage structure.

22. The first, that all slates whatever, among the older rocks, are more or less metamorphic; and that all metamorphism implies the development of crystalline force. Neither the chiastolite in the slate of Skiddaw, nor the kyanite in that of St. Gothard, could have been formed without the exertion, through the whole body of the rock, of crystalline force, which, extracting some of its elements, necessarily modifies the structure of the rest. The second, that slate-quarries of commercial value, fortunately rare among beautiful mountains, owe their

utility to the unusual circumstance of cleaving, over the quarryable space, practically in one direction only. But such quarryable spaces extend only across a few fathoms of crag, and the entire mass of the slate mountains of the world is cloven, not in one, but in half a dozen directions, each separate and explicit; and requiring, for their production on the pressure theory, the application of half a dozen distinct pressures, of which none shall neutralize the effect of any other! That six applications of various pressures at various epochs, might produce six cross cleavages, may be conceived without unpardonable rashness, and conceded without perilous courtesy; but before pursuing the investigation of this hexfoiled subject, it would be well to ascertain whether the cleavage of any rock whatever does indeed accommodate itself to the calculable variations of a single pressure, applied at a single time.

23. Whenever a bed of rock is bent, the substance of it on the concave side must be compressed, and the substance of it on the convex side, expanded. The degree in which such change of structure must take place may be studied at ease in one's arm-chair with no more apparatus than a stick of sealing-wax and a candle; and as soon as I am shown a bent bed of any rock with distinct lamination on its concave side, traceably gradated into distinct crevassing on its convex one, I will admit without farther debate the connection of foliation with pressure.

24. In the meantime, the delicate experiments by the conduct of which Professor Tyndall brought his audiences into what he is pleased to call "contact with facts" (in older times we used to say 'grasp of facts'; modern science, for its own part, prefers, not unreasonably, the term 'contact,' expressive merely of occasional collision with them,) must remain inconclusive. But if in the course of his own various 'contact with facts' Professor Tyndall has ever come across a bed of slate squeezed between two pieces of glass—or anything like them—I will thank him for a description of the locality. All metamorphic slates have been subjected assuredly to heat —probably to pressure; but (unless they were merely the shaly portions of a stratified group) the pressure to which they have been subjected was that of an irregular mass of rock ejected in the midst of them, or driven fiercely against them; and their cleavage—so far as it is indeed produced by that pressure, must be such as the iron of a target shows round a shell;—and not at all representable by a film of candle-droppings.

25. It is further to be observed,—and not without increasing surprise and increasing doubt,—that the experiment was shown, on the first occasion, to explain the lamination of slate, and on the second, to explain the riband structure of ice. But there are no ribands in slate, and there is no lamination in ice. There are no regulated alternations of porous with solid substance in the one; and there are no constancies of fracture by

plane surfaces in the other; moreover—and this is to be chiefly noted,—slate lamination is always straight; glacier banding always bent. The structure of the pressed wax might possibly explain one or other of these phenomena; but could not possibly explain both, and does actually explain neither.

26. That the arrangement of rock substance into fissile folia does indeed take place in metamorphic aluminous masses under some manner of pressure, has, I believe, been established by the investigations both of Mr. Sorby and of Mr. Clifton Ward. But the reasons for continuity of parallel cleavage through great extents of variously contorted beds;—for its almost uniform assumption of a high angle;—for its as uniform non-occurrence in horizontal laminæ under vertical pressure, however vast;—for its total disregard of the forces causing upheaval of the beds;—and its mysteriously deceptive harmonies with the stratification, if only steep enough, of neighbouring sedimentary rocks, remain to this hour, not only unassigned, but unsought.

27. And it is difficult for me to understand either the contentment of geologists with this state of things, or the results on the mind of ingenuous learners, of the partial and more or less contradictory information hitherto obtainable on the subject. The section given in the two lower figures of Plate VII. was drawn for me, as I have already said, by my most affectionately and reverently remembered friend, Professor Phillips, of Oxford. It

goes through the entire crest of the Lake district from Lancaster to Carlisle, the first emergent rock-beds being those of mountain limestone, A to B, not steeply inclined, but lying unconformably on the steeply inclined flags and grit of Furness Fells, B to C. In the depression at C lies Coniston Lake; then follow the masses of Coniston Old Man and Scawfell, C to D, sinking to the basin of Derwentwater just after the junction, at Grange, of their volcanic ashes with the Skiddaw slate. Skiddaw himself, and Carrock Fell, rise between D and E; and above E, at Caldbeck, again the mountain limestone appears in unconformable bedding, declining under the Trias of the plain of Carlisle, at the northern extremity of which a few rippled lines do service for the waves of Solway.

28. The entire ranges of the greater mountains, it will be seen, are thus represented by Professor Phillips as consisting of more or less steeply inclined beds, parallel to those of the Furness shales; and traversed by occasional cleavages at an opposite angle. But in the section of the Geological Survey, already referred to, the beds parallel to the Furness shales reach only as far as Wetherlam, and the central mountains are represented as laid in horizontal or slightly basin-shaped swirls of ashes, traversed by ejected trap, and divided by no cleavages at all, except a few vertical ones indicative of the Tilberthwaite slate quarries.

29. I think it somewhat hard upon me, now that I am

sixty years old, and short of breath in going up hills, to have to compare, verify for myself, and reconcile as I may, these entirely adverse representations of the classical mountains of England:—no less than that I am left to carry forward, in my broken leisure, the experiments on viscous motion instituted by James Forbes thirty years ago. For the present, however, I choose Professor Phillips' section as far the most accurately representative of the general aspect of matters, to my present judgment; and hope, with Mr. Clifton Ward's good help, to give more detailed drawings of separate parts in the next volume of Deucalion.

30. I am prepared also to find Professor Phillips' drawing in many respects justifiable, by my own former studies of the cleavage structure of the central Alps, which, in all the cases I have examined, I found to be a distinctly crystalline lamination, sometimes contorted according to the rock's own humour, fantastically as Damascus steel; but presently afterwards assuming inconceivable consistency with the untroubled repose of the sedimentary masses into whose company it had been thrust. The junction of the contorted gneiss through which the gorge of Trient is cleft, with the micaceous marble on which the tower of Martigny is built, is a transition of this kind within reach of the least adventurous traveller; and the junction of the gneiss of the Montanvert with the porous limestone which underlies it, is certainly the most interesting, and the most easily explored, piece of rock-fellow-

ship in Europe. Yet the gneissitic lamination of the Montanvert has been attributed to stratification by one group of geologists, and to cleavage by another, ever since the valley of Chamouni was first heard of: and the only accurate drawings of the beds hitherto given are those published thirty years ago in 'Modern Painters.' I had hoped at the same time to contribute some mite of direct evidence to their elucidation, by sinking a gallery in the soft limestone under the gneiss, supposing the upper rock hard enough to form a safe roof; but a decomposing fragment fell, and so nearly ended the troubles, with the toil, of the old miner who was driving the tunnel, that I attempted no farther inquiries in that practical manner.

31. The narrow bed, curved like a sickle, and coloured vermilion, among the purple slate, in the uppermost section of Plate VII., is intended to represent the position of the singular band of quartzite and mottled schists, ("bunte schiefer,") which, on the authority of Studer's section at page 178 of his second volume, underlies, at least for some thousands of feet, the granite of the Jungfrau; and corresponds, in its relation to the uppermost cliff of that mountain, with the subjacence of the limestone of Les Tines to the aiguilles of Chamouni. I have coloured it vermilion in order to connect it in the student's mind with the notable conglomerates of the Black Forest, through which their underlying granites pass into the Trias; but the reversed position which it here assumes, and the relative dominance of the central mass of the

Bernese Alps, if given by Studer with fidelity, are certainly the first structural phenomena which the geologists of Germany should benevolently qualify themselves to explain to the summer society of Interlachen. The view of the Jungfrau from the Castle of Manfred is probably the most beautiful natural vision in Europe; but, for all that modern science can hitherto tell us, the construction of it is supernatural, and explicable only by the Witch of the Alps.

32. In the meantime I close this volume of Deucalion by noting firmly one or two letters of the cuneiform language in which the history of that scene has been written.

There are five conditions of rock cleavage which the student must accustom himself to recognize, and hold apart in his mind with perfect clearness, in all study of mountain form.

I. The Wave cleavage: that is to say, the condition of structure on a vast scale which has regulated the succession of summits. In almost all chains of mountains not volcanic, if seen from a rightly chosen point, some law of sequence will manifest itself in the arrangement of their eminences. On a small scale, the declining surges of pastoral mountain, from the summit of Helvellyn to the hills above Kendal, seen from any point giving a clear profile of them, on Wetherlam or the Old Man of Coniston, show a quite rhythmic, almost formal, order of ridged waves, with their steepest sides to the lowlands;

for which the cause must be sought in some internal structure of the rocks, utterly untraceable in close section. On vaster scale, the succession of the aiguilles of Chamouni, and of the great central aiguilles themselves, from the dome of Mont Blanc through the Jorasses, to the low peak of the aiguille de Trient, is again regulated by a harmonious law of alternate cleft and crest, which can be studied rightly only from the far-distant Jura.

The main directions of this vast mountain tendency might always be shown in a moderately good model of any given district, by merely colouring all slopes of ground inclined at a greater angle than thirty degrees, of some darker colour than the rest. No slope of talus can maintain itself at a higher angle than this, (compare 'Modern Painters,' vol. iv., p. 318;) and therefore, while the mathematical laws of curvature by aqueous denudation, which were first ascertained and systematized by Mr. Alfred Tylor, will be found assuredly to regulate or modify the disposition of masses reaching no steeper angle, the cliffs and banks which exceed it, brought into one abstracted group, will always display the action of the wave cleavage on the body of the yet resisting rocks.

33. II. The Structural cleavage.

This is essentially determined by the arrangement of the plates of mica in crystalline rocks, or—where the mica is obscurely formed, or replaced by other minerals —by the sinuosities of their quartz veins. Next to the actual bedding, it is the most important element of form in minor masses of crag; but in its influence on large contours, subordinate always to the two next following orders of cleavage.

34. III. The Asphodeline cleavage;—the detachment, that is to say, of curved masses of crag more or less concentric, like the coats of an onion. It is for the most part transverse to the structural cleavage, and forms rounded domes and bending billows of smooth contour, on the flanks of the great foliated mountains, which look exactly as if they had been worn for ages under some river of colossal strength. It is far and away the most important element of mountain form in granitic and metamorphic districts.

35. IV. The Frontal cleavage. This shows itself only on the steep escarpments of sedimentary rock, when the cliff has been produced in all probability by rending elevatory force. It occurs on the faces of nearly all the great precipices in Savoy, formed of Jura limestone, and has been in many cases mistaken for real bedding. I hold it one of the most fortunate chances attending the acquisition of Brantwood, that I have within three hundred yards of me, as I write, jutting from beneath my garden wall, a piece of crag knit out of the Furness shales, showing frontal cleavage of the most definite kind, and enabling me to examine the conditions of it as perfectly as I could at Bonneville or Annecy.

36. V. The Atomic cleavage.

This is the mechanical fracture of the rock under the

hammer, indicating the mode of coherence between its particles, irrespectively of their crystalline arrangements. The conchoidal fractures of flint and calcite, the raggedly vitreous fractures of quartz and corundum, and the earthy transverse fracture of clay slate, come under this general head. And supposing it proved that slaty lamination is indeed owing either to the lateral expansion of the mass under pressure, or to the filling of vacant pores in it by the flattening of particles, such a formation ought to be considered, logically, as the ultimate degree of fineness in the coherence of crushed substance; and not properly a 'structure.' I should call this, therefore, also an 'atomic' cleavage.

37. The more or less rectilinear divisions, known as 'joints,' and apparently owing merely to the desiccation or contraction of the rock, are not included in the above list of cleavages, which is limited strictly to the characters of separation induced either by arrangements of the crystalline elements, or by violence in the methods of rock elevation or sculpture.

38. If my life is spared, and my purposes hold, the second volume of Deucalion will contain such an account of the hills surrounding me in this district, as shall be, so far as it is carried, trustworthy down to the minutest details in the exposition of their first elements of mountain form. And I am even fond enough to hope that some of the youths of Oxford educated in its now established schools of Natural History and Art, may so securely and

consistently follow out such a piece of home study by the delineation of the greater mountains they are proud to climb, as to redeem, at last, the ingenious nineteenth century from the reproach of having fostered a mountaineering club, which was content to approve itself in competitive agilities, without knowing either how an aiguille stood, or how a glacier flowed; and a Geological Society, which discoursed with confidence on the catastrophes of chaos, and the processes of creation, without being able to tell a builder how a slate split, or a lapidary how a pebble was coloured.

# APPENDIX.

---

When I began Deucalion, one of the hopes chiefly connected with it was that of giving some account of the work done by the real masters and fathers of Geology. I must not conclude this first volume without making some reference, (more especially in relation to the subjects of inquiry touched upon in its last chapter,) to the modest life and intelligent labour of a most true pioneer in geological science, Jonathan Otley. Mr. Clifton Ward's sketch of the good guide's life, drawn up in 1877 for the Cumberland Association for the Advancement of Literature and Science, supplies me with the following particulars of it, deeply—as it seems to me—instructive and impressive.

He was born near Ambleside, at Nook House, in Loughrigg, January 19th, 1766. His father was a basket-maker ; and it is especially interesting to me, in connection with the resolved retention of Latin as one of the chief elements of education in the system I am arranging for St. George's schools, to find that the Westmoreland basket-maker was a good Latin scholar ; and united Oxford and Cambridge discipline for his son with one nobler than either, by making him study Latin and mathematics, while, till he was twenty five, he worked as his father's journeyman at his father's handicraft. "He also cleaned all the clocks and watches in the neighbourhood and showed himself very skilful in engraving upon copper-plates, seals and coin." In 1791 he moved to Keswick, and there lived sixty-five years, and died, ninety years old and upwards.

I find no notice in Mr. Ward's paper of the death of the father, to whose good sense and firmness the boy owed so much. There was yet a more woful reason for his leaving his birthplace. He was in love with a young woman named Anne Youdale, and had engraved their names together on a silver coin. But the village blacksmith, Mr. Bowness, was also a suitor for the maiden's hand ; and some years after, Jonathan's niece, Mrs. Wilson, asking him how it was that his name and Anne Youdale's were engraved together on the same coin, he replied, "Oh, the blacksmith beat me."* He never married, but took to mineralogy, watchmaking, and other consolatory pursuits, with mountain rambling—alike discursive and attentive. Let me not omit what thanks for friendly help and healthy stimulus to the earnest youth may be due to another honest Cumberland soul, --Mr. Crosthwaite. Otley was standing one day (before he removed to Keswick) outside the Crosthwaite Museum,† when he was accosted by its founder, and asked if he would sell a curious stick he held in his hand. Otley asked a shilling for it, the proprietor of the Museum stipulating to show him the collection over the bargain. From this time congenial tastes drew the two together as firm and staunch friends.

He lived all his life at Keswick, in lodgings,—recognized as "Jonathan Otley's, up the steps,"—paying from five shillings a week at first, to ten, in uttermost luxury; and being able to give account of his keep to a guinea, up to October 18, 1852,—namely, board and lodging for sixty-one years and one week, £1325; rent of room extra, fifty-six years, £164 10*s*. Total keep and roof overhead, for the sixty usefullest of his ninety years, £1489 10*s*.

Thus housed and fed, he became the friend, and often the teacher, of the leading scientific men of his day,—Dr. Dalton the chemist, Dr. Henry the chemist, Mr. Farey the engineer, Airy the Astronomer Royal, Professor Phillips of Oxford, and Professor Sedgwick of Cambridge. He was the first accurate describer and accurate map-maker of the Lake district; the founder of the geological divisions of its rocks,—which were accepted from him by Sedgwick, and are now finally confirmed;—and the first who clearly defined the separation between bedding, cleavage, and joint in rock,—hence my enforced notice of him, in this place. Mr. Ward's memoir gives examples of

---

* I doubt the orthography of the fickle maid's name, but all authority of antiquaries obliges me to distinguish it from that of the valley. I do so, however, still under protest—as if I were compelled to write Lord Lonsdale, 'Lownsdale,' or the Marquis of Tweeddale, 'Twaddle,' or the victorious blacksmith, 'Beauness.' The latter's family still retain the forge by Elter Water—an entirely distinct branch, I am told, from our blacksmiths of the Dale: *see* above, pp. 189, 190.

† In that same museum, my first collection of minerals—fifty specimens—total price, if I remember rightly, five shillings—was bought for me, by my father, of Mr. Crosthwaite. No subsequent possession has had so much influence on my life. I studied Turner at his own gallery, and in Mr. Windus's portfolios; but the little yellow bit of "copper ore from Coniston," and the "Garnets" (I never could see more than one!) from Borrowdale, were the beginning of science to me which never could have been otherwise acquired.

his correspondence with the men of science above named: both Phillips and Sedgwick referring always to him in any question touching Cumberland rocks, and becoming gradually his sincere and affectionate friends. Sedgwick sate by his death-bed.

I shall have frequent occasion to refer to his letters, and to avail myself of his work. But that work was chiefly crowned in the example he left—not of what is vulgarly praised as self-*help*, (for every noble spirit's watchword is "God us ayde")—but of the rarest of mortal virtues, self-*possession*. "In your patience, possess ye your souls."

I should have dwelt at greater length on the worthiness both of the tenure and the treasure, but for the bitterness of my conviction that the rage of modern vanity must destroy in our scientific schoolmen, alike the casket, and the possession.

# INDEX.

CPSIA information can be obtained at www.ICGtesting.com
Printed in the USA
LVOW04s1325230715

447361LV00026B/465/P